AMERICAN SCIENTISTS
AND NUCLEAR WEAPONS POLICY

American Scientists
and
Nuclear Weapons Policy

BY ROBERT GILPIN

PRINCETON, NEW JERSEY

PRINCETON UNIVERSITY PRESS

1962

To Jean

ACKNOWLEDGMENTS

In writing this book, I have benefited enormously from the criticisms and assistance of many individuals, especially social and natural scientists, and I would like to express my profound gratitude for their generous assistance. The credit for whatever merit this book may have is due in large measure to them; the responsibility for its demerits is mine alone.

This book has grown out of work which began as research for my doctoral dissertation in political science at the University of California at Berkeley. The advice and ideas of the members of my dissertation committee—Paul Seabury, Norman Jacobson, and A. Hunter Dupree—were invaluable in the preparation of that initial study and have continued to influence my thoughts in this field.

My appreciation of many of the problems discussed in this book was greatly enhanced by my experience as a member of the Congressional Internship Program of the American Political Science Association in 1959–1960. At that time I had the pleasure of serving on the staff of Congressman Jack Westland and that of Senator Albert Gore; I learned much from their respective administrative assistants, James Dolliver and William Allen.

I am grateful to Dean Don K. Price of the Graduate School of Public Administration, Harvard University, who invited me to spend a year in Cambridge as a Fellow of the School's Science and Public Policy Program 1960–1961. Not only was I thus afforded an opportunity to prepare this study for publication, but I benefited from the counsel of Dean Price and enjoyed the fellowship of others interested in many of the problems which are the concern of the book. Among such persons with whom I spent many hours discussing problems of mutual interest were Robert Toth, Nieman Fellow of the *New York Herald Tribune,* and Major Robert Gard, U. S. Army.

Furthermore, I am most appreciative of the support and assistance given to me in the final preparation of the manuscript by the Council for Atomic Age Studies of Columbia University. In particular, I am indebted to Christopher Wright, Executive Director of the Council, whose profound understanding of many matters discussed in this book made his advice of great benefit and to Professor William T. R. Fox, Council member and Director of the Institute of War and Peace Studies of Columbia University, who was a source of encouragement when my spirits were lagging.

The scope of this study—touching upon the realms of both politics and technology—and the absence of an adequate body of secondary literature necessitated my calling upon several other colleagues in the social sciences and upon many natural scientists for criticism and guidance. Particular thanks for criticisms of the manuscript are due to social scientists Nathan Glazer, Morton Halperin, Alfred Hotz, and Henry Rowen. The co-operation of the many natural scientists and public officials who read sections of the manuscript and gave me generous interviews has been invaluable; the knowledge and insights of these individuals have enabled me to correct many errors of fact and judgment which appeared in the earlier drafts of the manuscript. My gratitude to them is considerable although, in accordance with their own requests and my own judgment, they shall remain anonymous.

In closing, I would like to acknowledge a special indebtedness to my wife, Jean. As critic, editor, and intellectual companion she made intelligible what otherwise might have been incoherent and obscure. Where such deficiencies remain, there, no doubt, my will prevailed over hers and for this I make due apologies both to her and to the reader.

ROBERT GILPIN

March 1962

CONTENTS

AMERICAN SCIENTISTS
AND NUCLEAR WEAPONS POLICY

" . . . Scientists sometimes have been called too optimistic and naïve by social scientists. As a group, they are not lacking in idealism. Perhaps it was natural that many of us, recognizing from close at hand the significance of nuclear weapons, set out to advise the world that nuclear war was out of the question. To us, the data were unequivocal, the conclusions indisputable, and the course of action clear. We felt the world would quickly see this—and, seeing it, do something about it.

"The half-life of disillusionment varied from individual to individual. Few have changed their minds about nuclear war. But, many have become more sophisticated, if less idealistic. Much of what has been described as naïveté has rubbed off. But we should remember that idealism, happily, has not been limited to scientists. In the period following World War I, experienced statesmen, imperceptibly influenced by scientists, solemnly signed unrealistic treaties outlawing war. Perhaps sophisticated statesmen, aided by sophisticated scientists in an age of science, may be able to combine realism and idealism."

GLENN T. SEABORG, Chairman of the United States Atomic Energy Commission and 1951 Nobel Laureate in chemistry—Excerpt from Speech to AAAS, delivered December 1961.

CHAPTER ONE

THE SCIENTIST AS A POLITICAL ANIMAL

" . . . it is evident that the polis belongs to the class of things that exist by nature, and that man is by nature an animal intended to live in a polis. He who is without a polis, by reason of his own nature and not of some accident, is either a poor sort of being, or a being higher than man . . . "—

ARISTOTLE, *Politics*

ALTHOUGH the admonition of Robert Hooke to his fellow natural scientists that they should " . . . improve the knowledge of all natural things . . . by Experiment (not meddling with Divinity, Metaphysics, Moralls, Politicks, Grammar, Rhetoric, or Logick)"[1] has traditionally been accepted by the American scientific community, today natural scientists are "meddling" in all manner of "Moralls and Politicks." Many are found as political activists in various voluntary organizations and as advisors at the highest levels of the United States Government. Examples of scientists' political activities have been abundant in the period since World War II. In 1945 American scientists formed numerous political action organizations which successfully helped to lobby important legislation through Congress.[2] In 1956 many scientists organized in support of opposing candidates in the presidential election. And today scientists advise the President on many issues, negotiate international agreements, and hold television debates on subjects of national policy.

The benefits to society of the scientists' new political role

[1] Quoted in Don K. Price, "Organization of Science Here and Abroad," *Science,* Vol. 129, No. 3351, March 20, 1959, p. 759.
[2] The most important piece of legislation for which scientist organizations such as the Federation of American Scientists lobbied was the act to establish the Atomic Energy Commission.

have been enormous. Creative, dedicated, and selfless minds have been brought into the realm of public affairs. There can be little doubt that without this new participation of scientists in political life the United States could not meet the increasingly difficult problems it faces as the world undergoes the twentieth century scientific revolution. Achievement of a truly effective utilization of this creative and dedicated talent should therefore be a major goal for American political leadership.

Since 1945, scientists have persistently brought their intelligence and energies to bear on the problem of national survival in the nuclear age and have assumed the responsibility of educating mankind to the dangers inherent in this new era. Unfortunately the scientists' conscientious concern over national survival has often been accompanied by a failure of American political leadership either to recognize the nature of the scientists' new political activity or to utilize it properly. While scientists have been given an increasingly important part in the formation of national policy on such questions as the decisions to build the hydrogen bomb, to develop tactical nuclear weapons, and to seek a nuclear test ban, political leaders have frequently failed to evaluate scientific advice effectively and to incorporate it into national policy.

Furthermore, the scientist's integration into political life has been inhibited by the scientist's ambiguous image of himself. At the same time that he resents the popular notion that he is somehow different from other people, the scientist believes that he carries with him into the political arena certain unique habits of mind which lend him advantage in understanding politics. Specifically, the scientist regards himself as being able to approach political issues with the same dispassionate, objective state of mind that he believes he displays in his scientific endeavors.

This equivocal self-image of the scientist was reflected in a colloquy between the scientist-interrogator, Ward Evans,[3] and

___

[3] Ward Evans, a member of the Personnel Security Board, was a professor of chemistry.

one scientist-witness, Norris Bradbury,[4] during the security trial of Robert Oppenheimer in 1954. Throughout much of the hearing, Evans demonstrated a sensitivity to the implication that scientists are somehow different from other people:

EVANS: Do you think that scientific men as a rule are rather peculiar individuals?

BRADBURY: When did I stop beating my wife?

GRAY:[5] Especially chemistry professors?

EVANS: No, physics professors.

BRADBURY: Scientists are human beings. I think as a class, because their basic task is concerned with the exploration of the facts of nature, understanding, this is a quality of mind philosophy—a scientist wants to know. He wants to know correctly and truthfully and precisely. By this token it seems to me he is more likely than not to be interested in a number of fields, but to be interested in them from the point of view of exploration. What is in them? What do they have to offer? What is their truth? I think this degree of flexibility of approach, of interest, of curiosity about facts, about systems, about life, is an essential ingredient to a man who is going to be a successful research scientist. If he does not have this underlying curiosity, willingness to look into things, wish and desire to look into things, I do not think he will be either a good or not certainly a great scientist.

Therefore, I think you are likely to find among people who have imaginative minds in the scientific field, individuals who are also willing, eager to look at a number of other fields with the same type of interest, willingness to examine, to be convinced and without *a priori* convictions as to rightness or wrongness, that this constant or this or that curve or this or that function is fatal.

[4] Norris Bradbury replaced Oppenheimer as technical director of the Los Alamos Weapons Laboratory.

[5] Gordon Gray was the chairman of the Personnel Security Board which investigated the Oppenheimer case; the Board is often referred to as the "Gray Board."

I think the same sort of willingness to explore other areas of human activity is probably characteristic. If this makes them peculiar, I think it is probably a desirable peculiarity.[6]

In the same sense in which Bradbury has characterized his "peculiarities" as a scientist this study will examine the "peculiarities" of the scientist as a political animal, the significance of these "peculiarities" for policy-making, and the changes required in the government's utilization of scientists if scientists are to function there for maximum social benefit. The basic purpose of the study, then, is to achieve at least a partial understanding of the scientist as a participant in political life. In what ways is he fitted for political activities? In what ways is he not? What are his unique contributions to national policy-making? What are his political goals? Such questions as these will be investigated in this book. The history of the intra-scientific conflict over nuclear weapons and its effect on the participation of scientists in the formulation of American nuclear policy will be examined. Such a subject has an intrinsic value in its own right because national survival depends in great part upon the soundness of scientists' advice on nuclear weapons policy. What has been the content of that advice? How has it been affected by the intra-scientific conflict over nuclear weapons? What are the implications of this conflict for effective decision-making in the scientific age?

It is hoped that this study will contribute not only to the immediate question of how scientists can be more effectively utilized in the formulation of a sound policy toward nuclear weapons but also to that of integrating scientific advice into the effort to solve the vast range of problems created by the advance of science. To what extent is the American experience in the formulation of policy toward nuclear weapons relevant to effective decision-making in other areas where science and politics merge?

[6] U.S. Atomic Energy Commission, *In the Matter of J. Robert Oppenheimer,* U.S.G.P.O., 1954, p. 491.

In order to answer questions about the scientist as a participant in political life, it is necessary first to ask: Who ought to be included in the definition of "scientist"? With what fraction of the population covered by the definition of "scientist" will the study deal? Are the "vocal" scientists—those who write and make speeches on nuclear weapons and whose ideas are available to the researcher—really representative of the scientists who are important for the subject of this study?

By "scientists," we shall mean those persons who are scientists according to the public image. The application of this term is not restricted only to persons whose task it is to discover new knowledge of the physical world; this definition also includes other individuals who by training or official position have spoken on public issues as representatives of the scientific community. Thus the definition used here includes an engineering-trained science administrator such as James Killian who served as President Eisenhower's Special Assistant for Science and Technology and a science-trained political analyst such as Ralph Lapp.

Although this definition of scientist is by necessity a broad one, the number of scientists actually of interest to this study is relatively small. The persons who as scientists have had an influential role in the determination of American nuclear policy have been a small fraction of the total scientific community. They include scientists such as Edward Teller and Robert Oppenheimer who have held important advisory positions in the councils of government as well as scientists like Linus Pauling and Eugene Rabinowitch who have been influential in the public intra-scientific debate over nuclear weapons.

This is, then, a study of the *effective* scientific opinion on American nuclear weapons policy, i.e., of the ideas of those scientists who have directly or indirectly influenced American policy toward nuclear weapons. As such it is unnecessary to consider whether or not the ideas of the scientists discussed here are really representative of all scientists. However, it is pertinent to question whether or not the individual "vocal" scientists do reflect the effective public opinion of scientists with respect to nuclear weapons. This is to ask: Are the scientists whom the writer has chosen to study and to quote representative of those scientists who have been most influential in the determination of American policy toward nuclear weapons? To what extent, for example, do the statements of a particular scientist in the *Bulletin of the Atomic Scientists* represent the views of his colleagues who as advisors to the government are influencing American nuclear policy? Furthermore, do the public statements of scientists who advise political leadership coincide with the advice privately given to governmental officials?

While it would require an exhaustive methodological examination to prove conclusively that the vocal scientists are indeed representative of effective scientific opinion, the writer feels that they are and believes his assumption to be justifiable for the following reasons:

1. The Oppenheimer security hearings released in 1954 reveal a similarity of opinion between the spokesmen for various scientific positions in the government and vocal spokesmen for the same positions writing in journals such as the *Bulletin of the Atomic Scientists*. The hearings reveal as well that, in cautiously written articles stripped of security information, the scientific advisors to the government were saying much the same thing to their scientific colleagues and to the public that they were saying to the government. Since some of these same persons are still advising the government and speaking publicly, the writer believes this situation remains largely unchanged today.

2. The intimacy between governmental advisors and vocal scientists unassociated with the government is frequently such that the vocal outsider is generally aware of the broad policy issues being debated inside the government and of the positions taken by fellow scientists. Again the transcript of the Oppenheimer security hearings reveals how well informed the out-

siders were on the issues of the H-bomb struggle then taking place within the bureaucracy.

3. The writer has become very much impressed by the uniformity and consistency of scientific political opinion. There are common assumptions and notions which characterize each set of scientists discussed in this work. Fortunately, government scientific advisors have been sufficiently vocal over the years to make possible identification of the ideas and assumptions they share with the more vocal outsiders. As a consequence, it is usually possible to know the general attitudes of the insiders on a subject by knowing what particular outsiders are saying. Again the transcript of the Oppenheimer security hearings confirms this situation.

4. The writer's contacts and interviews with insiders and outsiders among the scientists support his belief that the vocal scientists more often that not speak for *effective* scientific opinion. The extent to which, and the occasions upon which, this has been true are the subject matter of this study. The vocal scientists cited here seldom speak solely for themselves but, rather, they tacitly express ideas which they share with many silent yet frequently influential colleagues.

In summary, therefore, this is a study of the politically relevant ideas, attitudes, and behavior of those scientists who have been influential in the formulation of American policy toward nuclear weapons. As such, it focuses on a novel phenomenon in American political life, the emergence of scientists as men of political power.

The Political Rise of Scientists

The active participation of scientists in political life is a rather recent development, having taken place to a significant degree only within the last two decades. Although individual scientists and even scientific organizations have previously, upon occasion, taken part in politics and although governments have depended heavily upon advice and assistance from scientists in times of crisis, the extent of the present-day political activities of scientists suggests something of a revolution in the relationship of the scientist to politics. Never before has the participation of scientists in the determination of public policy been as pervasive or as important as it is today.

Two examples from World War I serve to illustrate the mutual indifference of scientist and government characteristic of the earlier period. In England, at the height of the war, Ernest, later Lord, Rutherford rejected the call to do research for the armed services because his independent experiments suggested a discovery which, in his words, would "eclipse the importance of the war," namely "the artificial disintegration of the atom." And in the United States at this same time, Thomas Alva Edison, who had the responsibility of mobilizing civilian science and technology for the war effort, displayed a similar indifference with his suggestion to the Navy that it should bring into the war effort at least *one* physicist in case it became necessary to "calculate something."

By 1939 the train of events emanating from Rutherford's work and the parallel developments in other branches of science were such that World War II quickly became a "calculators' " war. Out of the scientists' laboratories came microwave radar, the proximity fuse, and the atomic bomb; and for the first time in history, weaponry developed during the course of a war became a significant factor in its outcome. Never again would or could a government relegate the scientist to a secondary position. Scientific research had become a major element in national power.

It is particularly noteworthy that the scientist also became important to government in a capacity other than just that of laboratory researcher. Scientists were equally as responsible for the initiation of projects which would lead to the development of new weapons as they were for the research execution of the projects. Thus it was a group of scientists which suggested to the United States that it undertake the development of the atomic bomb. The success of the resultant Manhattan District Project in creating that bomb ushered in a new era in history—the age of the atom.

The world is now in the seventeenth year of the atomic age, and during these past seventeen years three more nations have joined the United States as nuclear powers. Today other nations stand or will soon be standing at the threshold of joining the "atomic club." At the same time the magnitude of possible weapons has increased many thousandfold from the 19 kiloton fission weapon used at Hiroshima to the Soviet 50-plus megaton fusion explosion of October 1961.[7] The threat of these weapons has also been multiplied by developments in other areas which have made it possible for the weapons to be delivered anywhere on earth within minutes of their launching. As Robert Oppenheimer has described it, this situation provides our globe with "a very brittle kind of stability."[8]

The recognition by the United States Government of this world situation has prompted it to engage scientists as advisors at all levels of its activities. No development in this area has been more important than the scientists' new relationship to the Chief Executive. Although Americans today accept without question the notion that the President must have scientific advisors, this innovation in government appeared late in the struggle for technical-scientific supremacy among nations.

During World War II, the President did have a scientific advisor in the person of Vannevar Bush, Director of the Office of Scientific Research and Development (OSRD). Regrettably, the postwar dissolution of OSRD eliminated any direct source of scientific advice to the President even though during the war years science had become recognized as a major national resource. Bush made science an active partner with the military in the prosecution of the war but, until recently, science has had no comparable relationship with the government in the postwar period.

During the Truman Administration, however, the voice of science in the executive branch of government was maintained

---

[7] A kiloton explosion is equivalent to that of 1000 tons of TNT; a megaton to one million tons of TNT.

[8] As quoted in Ralph Lapp, *Atoms and People,* Harpers, 1956, p. 180.

as an unanticipated consequence of the atomic energy legislation. When Congress created the Atomic Energy Commission (AEC) in 1946, it established by statute a General Advisory Committee (GAC) to advise the AEC. The prestige of the atomic scientists appointed to the GAC, and especially that of its chairman, Robert Oppenheimer, who had been scientific director of the Los Alamos weapons laboratory, gave the Committee a position of prominence and a voice of authority. The generally prevailing ignorance concerning atomic energy and the cloud of secrecy which blanketed the subject combined to give the GAC a large part in determining AEC policy in the early postwar years. And, since United States military planning became very dependent upon atomic weapons during these years, the GAC found that its role as advisor to the AEC became more and more significant in its effect on national security policy.

However, in 1950, the advice of the GAC was overruled in the decision to develop the hydrogen bomb, and the GAC consequently declined in the influence it exerted on policy. Contrary to popular impressions, however, the influence of other scientists within the Administration did not decline. As a matter of fact, with the decision to build the hydrogen bomb and the decision to remobilize after the outbreak of the Korean War in 1950, the dependence of the government upon scientific advice increased. The Korean War created an urgent need for new weapons to counter Communist aggression because very little had been done to create new alternative military capabilities since the development of the atomic bomb.

At the prodding of scientists within the advisory machinery of the government, steps were taken to bring about closer integration of scientific advice with military planning. A Science Advisory Committee was established in the Office of Defense Mobilization and summer study projects were undertaken in various scientific-military fields such as underwater and land warfare. The recommendations which resulted from the deliberations of these study projects of scientists and military men produced a profound change in American military posture.

Despite the rising tide of McCarthyism which alienated many scientists from the government, the influence of scientists in government continued to rise in the early 1950's. At this time the informal and personal political influence on policy of the opinions of particular scientists was as significant as the structural changes to provide scientific advice within the government.

With the support of AEC Chairman Lewis Strauss, the influence of Edward Teller, who had been the major contributor to the development of the hydrogen bomb as well as the victor in the struggle over the decision to develop it, became strong in the Administration. In fact, at this point a major complaint from critics of the Administration was that the influence of scientists, or at least of this particular scientist, was too great in the determination of American nuclear policy. Whereas Edward Teller received a sympathetic hearing at the highest levels of the Administration, his scientific opponents were able to voice their arguments only at the lower level of the General Advisory Committee and in the relatively ineffective Science Advisory Committee which was still in the Office of Defense Mobilization. Since it was believed by his critics that Teller represented only a minority of scientific opinion, many scientists were dissatisfied with a situation which provided poor access to the White House for what they considered to be the majority view among scientists on issues of national policy.

But by 1955 events were in progress which would soon bring to the top echelon of the government scientists with views on nuclear weapons policy more representative of the scientific community as a whole. The thaw in the Cold War, the decline in McCarthyism, and the continued success of Soviet science were stimulating moves for a major change in science-government relations.

Above all else, it was the appearance of the Soviet Sputnik

on October 4, 1957 which caused a sweeping change in the relationship of science and government. By far the most important response to Sputnik—and there were many—was the appointment by President Eisenhower of James Killian, President of the Massachusetts Institute of Technology, as his Special Assistant for Science and Technology. At the same time the previously subordinate Science Advisory Committee was elevated to the White House level to serve as an advisory body to the President and his scientific advisor; approximately a year later, the Federal Council for Science and Technology was established. The Federal Council is composed of the heads of scientific agencies in the government while the President's Science Advisory Committee is composed of private citizens.

These innovations represent official recognition of the changed relationship between government and science in recent years. Appropriately the scientific community, or at least the larger part of it, now has the continuous and direct access to the President of the United States which it previously possessed only during the crucial years of World War II, and the President is now enabled to utilize the resources of American science more fully and regularly than ever before. Through this functional change within the executive branch of the government as well as many other developments including scientists' own private ventures into politics, scientists are today well established participants in the political process.

In particular, the scientist has become important in the formulation of American national security policy such as that dealing with nuclear weapons. Since 1945 the scientist's advice on the feasibility of policies has become an increasingly important factor in policy formation. Is it feasible to arrest the further development of nuclear weapons by a nuclear test ban? Is it feasible to develop a hydrogen bomb or low yield tactical nuclear weapons? The answers which scientists give to this type of question become significant ingredients in decision-making and have important ramifications for national survival.

The Political Nature of Scientists' Advice

The advice of experts to the policy-maker, including that of scientists, is seldom if ever solely technical. While the advice generally appears quite technical, careful analysis of its substance will reveal it to be political in nature. This is to say that even though the expert may present his advice in terms of the technical *what is,* the advice may be important politically because explicit or implicit in the reported technical data are numerous non-technical assumptions including political assumptions concerning *what ought to be done.* These non-technical assumptions influence and sometimes determine the problems selected by the expert for emphasis, the facts he believes relevant, and the implications he may draw from the selected facts for public policy. As a consequence the non-technical assumptions of the expert which influence his technical advice may be important factors in the formation of public policy.

Although in most cases these non-technical assumptions are provided for the expert by political leadership or are obvious enough in the expert's advice to be available for review by the person being advised, there are important exceptions to this situation. The degree to which the expert himself implicitly supplies the political assumptions underlying his advice depends upon a number of factors. It depends first of all on the level of the governmental hierarchy at which the expert advises; the higher the level, the more likely it is that the expert will be important not only in providing technical data but in participating in the selection of the political goals to be achieved. Similarly, as the technical complexity of the subject matter and the political indecisiveness of the persons advised increase, the expert's role in policy determination is also enlarged and his own non-technical assumptions become important ingredients in policy formation.

As a matter of fact, the situation in which the scientist-advisor to political leadership finds himself makes inevitable the introduction of non-technical views into his advice. This inevitability arises out of the vast gap in communications between the technical expert and the political generalist. Whereas the knowledge possessed by the expert is complex, abstruse, and often quantitative, the political decision-maker pressed by many problems requires qualitative advice: Ought he or ought he not to make a particular decision?

Under these circumstances, if the expert's knowledge is to be meaningful to the decision-maker, the expert himself must make it meaningful. He must select those facts which he believes are relevant to the problem under consideration and interpret their significance for public policy. Thus it is inevitable that the scientist's advice on technical matters be based in part on the scientist's non-technical assumptions about the problems at hand.

In the area of national policy toward nuclear weapons, all the conditions exist that permit and even encourage the expert to supply his own non-technical assumptions concerning policy goals and thus to exert a strong influence on the formulation of policy. Firstly, the scientists advise at the highest level of the governmental hierarchy. Secondly, the subject matter of atomic weaponry is highly complex; and thirdly, American political leadership has been extremely indecisive over the goals of its nuclear policy.

That type of scientist-expert advice with which this study deals occupies a middle category between the realm of science, or *what is,* and the realm of policy, or *what is to be done;* and therefore we find the scientist who advises on nuclear weapons inevitably combining his scientific and his political judgment. The scientific advisor to the President or to his top aides does not simply provide the advisee with an objective analysis of the facts but also with his judgment concerning *what is to be done,* a judgment based on the scientific facts as seen and interpreted by the expert. The necessity for such a combination of technical and political judgment gives to the scientist-advisor his special status.

This characterization of the expert's advice as part technical and part political is well recognized in some areas of public policy and is only just becoming appreciated in others. It is now commonplace, for example, to view with a critical eye the advice of experts from business or labor who appear before a Congressional committee. Also, it is increasingly appreciated that there is no purely military advice; President Eisenhower himself typified this new attitude with his reference to "parochial generals." As a matter of fact, concern resulting from this recognition was partially responsible for President Eisenhower's elevation of the Science Advisory Committee to the White House level where it would provide him with an alternative source of advice to that of the military and of other special interests in the government.

Despite the fact that scientific advice has thus been purposely sought as a balance to advice from other sources which are recognized as prejudiced and limited in their point of view, scientific advice has not in general been evaluated in similarly critical terms. On the contrary, the history of American nuclear policy leads one to conclude that both scientists and political leaders have acted as if it were possible to make a clear delineation in policy formulation between the political and the technical realms. And, according to this generally accepted view, the task of the scientist is to supply only the objective, technical data which the decision-maker can then combine with the non-technical ingredients of policy supplied by himself or some other expert.

This simplistic view of the scientist's role as advisor has created expectations which the scientist cannot fulfill. He is expected by political leadership, fellow scientists, and his own conscience to render only objective technical advice. As a consequence of this failure to appreciate the necessary intertwining of the technical and political, scientists have been assigned many apparently "technical" tasks whose performance has required a political skill far beyond their competence; scientists and political leaders have failed to realize the nature of the non-technical assumptions underlying the scientist's advice; and scientists have charged one another with intellectual dishonesty when they have disagreed strongly with advice which has been given.

The scientist's own failure to appreciate the complexities of the special role of the scientist as advisor has led him to apply the same criteria in evaluating advice to the government that he applies in evaluating the professional performance of colleagues. With few exceptions the scientist expects the advice of fellow scientists to be true and objective in the same sense that any scientific statement is true. He forgets that the scientist as advisor is not so much seeking the truth as he is seeking wisdom. The advisor's task is to combine his technical expertise with his overall understanding and good sense. He has moved out of the realm where the scientific method and the scientist's canons of criticism are the most relevant guides for judgment. His primary responsibility as an advisor is to be wise and of this neither scientists nor any other group have standards of evaluation.

Although the following statement from Hans Bethe, a former member of the President's Science Advisory Committee (PSAC), indicates partial recognition of the fact that the advice of scientists is not purely factual, even it fails to appreciate the real problem. Bethe writes: "The only other group of people [beside the military] who are fully informed [on national defense] are the scientists working on the weapons. Since they don't have, *a priori,* a professional interest one way or the other, they should be able to consider non-military factors, political as well as ethical ones. Their opinion is therefore valuable in arriving at a balanced decision."[9]

Yet, even though one can easily agree with Bethe on the valuable role which the scientist has to play in the formulation of national security policy, it must be pointed out that scientists often fail to appreciate how their own political dis-

[9] Hans Bethe, "Review of Robert Jungk's *Brighter Than a Thousand Suns,*" *Bull. Atom. Sci.,* Vol. 14, No. 10, December 1958, p. 428.

positions may affect their advice. Nor do they always realize the limits on their own competence to consider the political aspects of a policy question. Just as the military officer may have his professional interest which limits the wisdom of his advice, the scientist may be limited by his commitment to a particular course for American nuclear policy. The failure of scientists and of others to apply to the scientist-advisor standards similar to those applied to other advisors or to view him as a wisdom seeker rather than as a spokesman for scientific truth has been a constant source of recrimination and antagonism within the scientific community. Apparent scientific differences among scientists which have actually been differences over the political significance of agreed-upon scientific facts have resulted in charges and counter-charges of intellectual dishonesty. Upon close examination, however, one discovers that no scientist in the bitter history of the intra-scientific conflict over nuclear weapons policy has actually distorted the scientific facts and that the scientists at all times have been in general agreement on the facts even though their conclusions have frequently been radically different from one another. However, scientists have on occasion selected, emphasized, and interpreted those specific scientific facts which they have believed to support their own political positions on the wisest course for American nuclear policy.

The scientist is, of course, aware of the danger that his personal views may influence his presentation to the public and its officials of advice on technical matters. However, he believes that he can effectively guard against this possibility and can keep separate his presentation of the technical facts from his personal views on the implications of the facts for public policy. Unfortunately, the scientist generally fails to see that such separation is impossible under the circumstances which characterize his participation in government. Indeed it must be noted that the conflict between the scientist's belief in his own objectivity and his actual role as advisor is a subtle one.

This conflict is best understood in terms of the advice given to literary critics by D. H. Lawrence in the preface to his *Studies in Classic American Literature*[10] where he states that the task of the critic is to compare the "tale's moral" with the "author's moral."[11] Lawrence points out how often the moral which the author says he wishes to make is contradicted by the moral exemplified in the author's use of character, description, and narrative. The same inconsistency between profession and performance is often true of the scientist as advisor.

Time after time a scientist will preface a statement with the assertion that his intention is only to make an objective presentation of the facts and that he will leave to his audience the task of drawing out the implications of these facts for public policy. However, with few exceptions in this writer's experience, the conclusion or moral which the scientist himself has reached becomes very obvious in the scientist's narrative. In reading many "technical" presentations by scientists on social and political problems one is reminded of the following passage from the writings of the eighteenth century philosopher, David Hume: "In every system of morality, which I have hitherto met with, I have always remark'd, that the author proceeds for some time in the ordinary way of reasoning . . . when of a sudden I am surpriz'd to find, that instead of the usual copulations of propositions, *is* and *is not*, I meet with no proposition that is not connected with an *ought*, or an *ought not*. This change is imperceptible, but is, however, of the last consequence."[12]

An excellent example of the contrast between the scientist's stated "moral" and his "tale's moral" is found in a study entitled *Fallout*, edited by John Fowler. The moral of the study according to Fowler "is to inform, not to mold opinion" and "to present to the public the raw materials from which

---

[10] Doubleday, 1951. The writer is indebted to Norman Jacobson for first calling Lawrence's point to his attention.

[11] *ibid.*, p. 13.

[12] David Hume, *Treatise of Human Nature*, L. A. Selby-Bigge, ed., Oxford, 1st ed. 1888, reprinted 1958, Book III, Part I, Sect. 1, p. 469.

this opinion can be forged is both the privilege and the duty of the scientist."[13] This objective presentation of the facts and only this is to be the writer's moral. However, by the end of the book, the tale's moral has asserted itself. Its conclusion is a passionate argument for disarmament including a nuclear test ban. "The road to disarmament is certainly not easy," the author tells us, but he maintains, quoting Bertrand Russell, it is more "ennobling and splendid . . . [than] perfecting of weapons of man's destruction."[14] Perhaps this is so, but the statement of these sentiments hardly constitutes an objective presentation of facts without any intention to mold opinion.

In a manner similar to this, the scientist's "technical" presentation very often unwittingly becomes transformed into political advice. A paper or document whose stated purpose is to lay out the relevant scientific facts frequently concludes with a statement concerning what ought to be done; and, as he fails to realize that he has reached his conclusion concerning *what ought to be done* on the basis of his own normative views, the scientist on occasion believes that he has found an objective solution to the political problem under consideration, a solution with which everyone will agree.

This study will show that much of the history of the politics of scientists since 1945 has been a search for such objective solutions to political problems. The scientists have viewed themselves as searchers for solutions based on facts and therefore acceptable to all; they see themselves as discovering the truth and educating the world to it. Scientists reason that once others have been educated to the facts, they will find the solution of a particular problem as obvious as do the scientists themselves.

[13] John Fowler, ed., *Fallout,* Basic Books, 1960, pp. 9–10. As a matter of fact even the title is polemical; it gives the impression that the issue at stake in the debate over testing is fallout and not the relationship of testing to international peace.

[14] *ibid.,* pp. 187–88. Again the writer would like to emphasize that he does not argue against expression of normative views by scientists, but does believe that such expressions should be recognized and presented as such.

The fact that the scientist can seldom, if ever, free himself from his underlying non-technical assumptions is quite understandable although seldom appreciated. Indeed, even in scientific research where rigorous conditions exist to encourage objectivity, it is impossible for the scientist to free himself from his values and his implicit assumptions of a non-technical or non-scientific nature. The values of the scientist are an integral part of his research; these values affect the problems he selects for study, the facts which he believes are relevant, and the implications or hypotheses he draws from the facts.[15]

Truly, it can be said that without the scientist's emotional and normative commitments there could be no science. They provide the impetus for the dynamism of scientific research. "The notion that a scientist is a cool, impartial, detached individual is," James Conant tells us, "of course, absurd. The vehemence of conviction, the pride of authorship burn as fiercely among scientists as among any creative workers. Indeed, if they did not, there would be no advance in science."[16]

Similarly, without the commitment of the scientist to certain political goals, a high level of competence, vitality, and dedication could not exist among the government's scientific advisors. However, within the advisory mechanisms of government, there are to be found no rigorous conditions similar to those imposed on the scientist in his laboratory which make explicit the scientist's underlying assumptions. Thus Conant's caution that " . . . this emotional attachment to one's own point of view is particularly insidious in science because it is so easy for the proponent of a project to clothe his convictions in technical language"[17] becomes especially meaningful when

[15] The importance of the scientist's values to his science is discussed in Bernard Barber, *Science and the Social Order,* Glencoe, Illinois: The Free Press, 1952, pp. 84–100.

[16] James Conant, *Modern Science and Modern Man,* Doubleday, 1952, p. 114. Conant's point has been well documented by Robert Merton, "Priorities in Scientific Discovery: A Chapter in the Sociology of Science," *American Sociological Review,* Vol. 22, No. 6, December 1957, pp. 635–59.

[17] Conant, *op.cit.,* p. 114.

applied to the scientist in government. Without accepting Conant's possible implication that this clothing of personal convictions in technical language is deliberate, it should at least be understood that the scientist as expert advisor operates as much on the basis of his non-technical assumptions as on his command of a body of scientific knowledge.

What then are the non-technical assumptions which underlie the scientist's advice in the area of national policy toward nuclear weapons? In the first place, they define the scientist's image of his role as policy advisor; specifically, the scientist believes that he has a social responsibility to assist society in the solution of its problems. Secondly, scientists share some political attitudes which are significant for the content of their advice. And thirdly, there are matters upon which groupings of scientists hold conflicting political assumptions. Because each of these assumptions affects the scientist's advice, they must all be considered.

The Scientists' Sense of Social Responsibility

Since the Second World War increasing numbers of American scientists have chosen to follow what Max Weber has called an ethics of responsibility.[18] They have rejected equally the notion that the scientist's allegiance to higher ethical ends prohibits his service to the state in the realm of military research and the notion that the scientist ought to exclude his own ethical beliefs from this work. They have thus chosen to serve the state while at the same time assuming a responsibility to influence state policy along lines believed to be beneficial to mankind. In substance, scientists increasingly believe that they have a responsibility for the social implications of their scientific activities and that they can no longer leave to society

---

[18] The person, according to Weber, who follows an ethics of responsibility accepts the "ethical paradox" that in politics one must often utilize evil means such as violence to achieve good ends. H. H. Gerth and C. Wright Mills, *From Max Weber: Essays in Sociology*, New York: Oxford University Press, 1958, p. 121.

alone the task of assuring that scientific advance will be beneficial to mankind.

In a statement to which many American scientists would subscribe, Hans Bethe expressed his concept of the scientist's ethical or social responsibility. Disagreement undoubtedly would exist with respect to the ethical or political ends which the individual scientist desires to achieve and on the means to be employed but not on the basic question of the scientist's responsibility to his fellow man. Bethe stated:

"In order to fulfill this function of contributing to the decision-making process, scientists (at least some of them) . . . must be willing to work on weapons. They must do this also because our present struggle is (fortunately) not carried on in actual warfare which has become an absurdity, but in technical development for a potential war which nobody expects to come. The scientists must preserve the precarious balance of armament which would make it disastrous for either side to start a war. Only then can we argue for and embark on more constructive ventures like disarmament and international cooperation which may eventually lead to a more definite peace."[19]

Bethe's statement does not represent merely those scientists on one side of the intra-scientific conflict over nuclear weapons; Edward Teller, who represents a political position opposite to Bethe's, has also made much the same point concerning the scientist's social responsibility. "We," Teller says to his fellow scientists, "have two clear-cut duties: to work on atomic energy under our present administration and to work for a world government which alone can give us freedom and peace. It seems difficult to take on these responsibilities. To take on less, I believe, is impossible."[20]

The scientist's sense of social responsibility has risen largely out of the conflict between the scientist's image of himself as

[19] Hans Bethe, "Review of Robert Jungk's *Brighter Than a Thousand Suns,*" *Bull. Atom. Sci.,* Vol. 14, No. 10, December 1958, p. 428.

[20] Edward Teller, "Atomic Scientists Have Two Responsibilities," *Bull. Atom. Sci.,* Vol. 3, No. 12, December 1947, p. 356.

the disinterested searcher after truth and society's rising demands of him. This conflict was and is especially acute when the scientist is called upon to engage in military research and development and thereby to run counter to his image of himself as a member of an international community committed to the advancement of knowledge either for its own sake or for the benefit of all mankind.[21]

Whereas the relationship of the scientist to the warmaking capabilities of the state has traditionally been a remote one, the Manhattan District Project and the other wartime projects heralded the closely knit relationship of the military services and scientific research and development which exists today. Scientists in those wartime projects initiated a military revolution which changed irrevocably the relationship of science and war. Unaware of the implications of their activity and possessed by the spirit of the discovery of the unknown, these scientists pursued their work willingly and at times joyfully; at first, in the days prior to the destruction of Hiroshima, many of the scientists feared that their years of wartime research and effort might not prove fruitful.

Among the many groups of scientists working in the Manhattan District Project, however, there were aggregations of men who pondered the moral and political consequences of their actions even before the actual explosion of the atomic bomb. With success, a realization of what they had created swept over the great majority of the scientists; a sense of concern, and in many cases, even of guilt engulfed the participants in the project. Years later Robert Oppenheimer spoke for himself and for many of his fellow physicists when he wrote, "in some sort of crude sense which no vulgarity, no humor, no

[21] Although the reality of the "international community of science" has been challenged by some writers, the point made here is that American scientists do believe such a community exists. For a contrary view see A. Hunter Dupree, "Influence of the Past: An Interpretation of Recent Developments in the Context of 200 Years of History," *Perspectives on Science and Government*, The *Annals* of the American Academy of Political and Social Science, Vol. 327, January 1960, pp. 19–26.

overstatement can quite extinguish, the physicists have known sin; and this is a knowledge which they cannot lose."[22]

In his statement, Oppenheimer expresses the anguish of the scientist caught between his commitment to the pursuit of truth wherever it might lead and his realization that the knowledge he discovers may become the cause of great misery and destruction to mankind. Scientists have sought in varying ways to resolve this inner conflict, and a pervasive sense of a social responsibility concerning the utilization of scientific knowledge has arisen within the scientific community. Even though the scientist's new sense of social responsibility has grown in part from his feelings of guilt, it would be quite wrong to equate the two. The scientist's new sense of social responsibility has deeper roots than simply a feeling of guilt for past and present acts. The causes of this change in attitude are to be discovered in the radical changes in the relationship between science and society.[23] The mutual impact of science and society has stimulated in the scientist a desire to assist society in the solution of the problems created by scientific advance and also to participate in government in order to moderate the effects of government policies upon science itself.

Traditionally the view of the scientist toward his responsibility has been that expressed by Percy Bridgman who has argued that the only responsibility of the scientist is to discover the truth and to disseminate it, that he has no responsibility concerning society's utilization of that knowledge. If, Bridgman argued, scientists are held responsible for the evils made possible by scientific discoveries, this will lead ultimately to the political control of science in the interest of protecting society. The responsibility for the utilization of science, according to Bridgman, lies solely with society.[24]

The amoralism of science, or ethical positivism, taught by

[22] Robert Oppenheimer, "Physics in the Contemporary World," *Bull. Atom. Sci.,* Vol. 4, No. 3, March 1948, p. 66.

[23] For an excellent treatment of this theme, see Barber, *op.cit.,* pp. 207–37.

[24] Percy Bridgman, "Scientists and Society," *Bull. Atom. Sci.,* Vol. 4, No. 3, March 1948, pp. 69–72.

Bridgman, was believed in the past to be justified by the assumption that scientific progress was the road to human perfection. In essence, positivism as it is used here may be viewed as an ethic which sought to separate science and politics in the interest of human progress as well as of the freedom of science. The freedom and amoralism of science could be supported by the thesis that science was an autonomous force working for man's welfare in contrast to the disruptive force of politics.

This implicit ethic of positivism was increasingly challenged by world events after 1935. The inhuman use to which science was put by Nazi Germany, the world's leading scientific nation, shocked many scientists.[25] Many of these scientists emigrated to the United States as refugees during the 1930's; these emigrés were among the scientists who suggested the atomic bomb to President Roosevelt, formed the core of the Manhattan District Project and, after the war, sensed along with their American colleagues the "sin" of which Oppenheimer has written.

Whereas positivism had been adequate for prescribing the indirect relationship between basic research and society in earlier days, for most scientists it became inapplicable in the period after 1940 when the scientists themselves applied their knowledge and discoveries to military technology. Even though the tenacity of the positivist view was exhibited by Irving Langmuir in 1945 when he argued that the atomic bomb was really an accident for which scientists were not responsible, the terrible destructiveness of the results of the application of science to weaponry truly eradicated forever the faith in automatic progress through science. The moral question of whether or not scientific advance and social progress were even compatible was raised along with the companion query of how they could be made compatible if they were no longer so.

This challenging moral problem was faced first and felt

[25] An excellent discussion of the shock this produced among the scientists in the 1930's is found in C. P. Snow, "The Age of Rutherford," *The Atlantic*, Vol. 202, No. 5, November 1958, pp. 76–81.

most deeply by the scientific membership of the Manhattan District Project, the actual creators of the atomic bomb. The Federation of Atomic Scientists which was founded by Manhattan District Project scientists charged itself as follows in the words of the preamble to its constitution: "The Federation of Atomic Scientists is founded to meet the increasingly apparent responsibility of scientists in promoting the welfare of mankind and the achievement of a stable world peace." Although the entire scientific community has become permeated by a similar commitment, the scientists who worked on the atomic bomb have been more concerned with the moral problem than have others. And in none of their actions has this new concern that the utilization of scientific knowledge be beneficial to society been more deeply demonstrated than in their efforts to influence national policy toward nuclear weapons.

This belief of scientists that they have a responsibility to assist society in the solution of its problems is the foundation underlying the scientists' advice to government. It accounts both for the increased interest of scientists in public policy and for their willingness to participate actively in political affairs. Most importantly of all, in the formulation of governmental policy it accounts in large part for the passion with which certain scientists have supported their views on several problems upon which they have advised the government.

The Scientists' Shared Political Attitudes

In the exercise of their social responsibility, American scientists have revealed some shared political attitudes toward international politics, nuclear weapons, and the like; in addition, as the next section will indicate, scientists have exhibited conflicting political attitudes. Both the attitudes which unite and those which divide scientists in their political activities influence their advice to decision-makers and have had profound implications for national policy toward nuclear weapons.

Scientists tend to believe that scientific advance is taking mankind into a new period of history where the old rules of the statesmen no longer apply. They believe further that the scientist has a special understanding of this emerging new world which his science is creating; they envision the development of a new set of political rules based on the facts of a truly scientific age. At the same time that "the mainstream of political events is still dominated by traditional thinking and by the inertia of established institutions," scientists do optimistically believe that "the development of science and technology is rapidly changing the realities of human existence" and is destroying the "historical concepts of international struggle for power."[26] Mankind thus for the first time has the opportunity to leave behind the world of power politics and war. Fortunately, the scientists believe, mankind can take advantage of this new opportunity because the "art and science of controlling war have for the first time shown signs of genuine promise."[27]

The political outlook of the scientist is that of the optimist. He has, in the words of C. P. Snow, "the future in his bones." He believes *a priori* that there is a solution to be found to every problem and he expects to find the solution to the problem of atomic weapons just as he expects to find the solution to a problem in physics; he rejects the notion that the problem of atomic weapons may admit of no *final* solution but may be a problem with which man must deal as best he can for the rest of his existence.

In the fulfillment of his sense of social responsibility the quest of the scientist is for certitude. Despite the fact that science may be defined as organized skepticism and that scientific discoveries are continuously destroying previously accepted certainties, the individual scientist is psychologically committed to the discovery of *the* truth. His is the search for

[26] "The Dawn of a New Decade," *Bull. Atom. Sci.*, Vol. 16, No. 1, January 1960, pp. 5–6.

[27] Gerald Holton, "Editor's Prefatory Note," *Daedalus* (*Arms Control* Issue), Vol. 89, No. 4, Fall 1960, p. 675.

the permanent; he seeks constantly to flee from the lawless and contingent to the world of law and predictability.

Not only does the scientist seek *the* solution to a political problem but, in the words of Jerome Wiesner, President Kennedy's Special Assistant for Science and Technology, the "scientist is impatient in finding a solution;" he cannot be satisfied with small step accommodations such as those that comprise the bulk of political action. As Wiesner has noted, only large measures, despite their high degree of risk and danger, seem to the scientist to be sufficient to solve our pressing political problems such as that of atomic weapons.[28] The world of practical compromise within which the politician or administrator must operate is not tolerable to the scientist. Problems with which one learns to live rather than which one learns to solve are unacceptable to him. In the scientist's view, the reason for failure must lie with man's methods of dealing with the problem and not with the nature of the problem itself.

The scientist's quest for certainty and his confidence that reason and new methods can solve man's problems are reinforced by his conviction that something which is theoretically possible is also most likely to be politically possible. He accepts the notion that as the social world is the creation of man himself it is within man's power to change those things which are contrary to the common interest of mankind, providing only that mankind is willing to use his reason. The responsibility of the scientist is therefore to educate mankind to the necessity of trying to resolve seemingly intractable problems.

This attitude of the scientist often leads to a theoretical-experimental approach to the solving of social problems. If a solution to, say, the nuclear arms race is theoretically sound in that it is technically feasible and appears to reconcile the divergent interests of mankind, according to this view, it ought not be ruled out *a priori* on grounds of political im-

[28] "Scientists and World Politics," *Perspectives for Mankind,* Television Broadcast on Channel 2, Boston, December 18, 1960.

practicability. Instead, the necessity of finding some solution should compel men to try alternative approaches until one which works is found. In this sense certain groups of scientists have considered the Baruch Plan and a nuclear test ban to be experiments by which to test their underlying notions on the solution to the nuclear arms race.

This theoretical-experimental approach to the problems of politics accounts in large measure for the resilience and persistence of the scientist's political views developed during the course of World War II. In the face of repeated frustration the politically active scientist remains convinced that his theories have never really been tested or that experiments which have failed, like the Baruch Plan, had not been properly formulated. Scientists tend to reason that sooner or later a method for solving the problem will be devised.

A very strong conviction of the scientists is that science is a force for peace, and in particular they believe that science has a special role in the solution to the problem of atomic energy. However, they do differ on their definitions of this role. One view is that scientists, with their highly cultivated ability to think creatively and objectively, can influence their respective nations to pursue the reasonable policies which would ultimately lead to peace. Eugene Rabinowitch, editor of the *Bulletin of the Atomic Scientists,* expressed this faith very well when he wrote: "Because of the similarity in outlook of scientists all over the world, their increased influence on the national policies of the different countries should increase the ease of international communication. Their greater than average capacity for abstraction and generalization will favor policies based on long-range, rational planning—policies in which the enlightened self-interest of individual nations or political systems is bound to become coordinated with the common well-being of mankind."[29]

A second view on the relationship of science and peace is

[29] Eugene Rabinowitch, "History's Challenge to Scientists," *Bull. Atom. Sci.,* Vol. 12, No. 7, September 1956, p. 239. For a discussion of sources of the scientist's faith in reason, see Barber, *op.cit.*

that scientific progress and cooperation can be a major factor toward reducing the tensions that cause war. George Kistiakowsky, who served as President Eisenhower's Special Assistant for Science and Technology, has expressed this view: "Science also provides a sometimes unique opportunity for cooperative endeavors that can contribute in a major way to the reduction of tension between nations and, more positively, to close relations between the United States and other countries."[30]

A third view is that science can create the military technology to maintain the peace. In an interesting presentation of this position, C. W. Sherwin, then Chief Scientist, United States Air Force, derided all "social-political devices, such as disarmament, federal world government," etc. as impractical means to achieve the preservation of world peace and stated that: "On the other hand, new scientific knowledge and a new kind of cheap and easily transportable energy, offer a previously unknown technical tool for the assault on the problem of preserving peace."[31] According to Sherwin such technical developments and not "preatomic political concepts" can deter the outbreak of war and thus be " . . . a method of forcing good (or more accurately, rational) behavior through the fear of self destruction."[32] In this view perpetual peace will be achieved when opposing nations are demonstrably capable of destroying each other and also are unable to prevent retaliation for their attacks on others.

The confidence of most scientists in the axioms that science is a force for peace and that scientists have an important part to play in the solution to the problem of atomic weapons is quite comprehensible. The scientist who compares his own successes and those of his colleagues in problem-solving with the failure of the world's statesmen comes to believe that he might well succeed with his methods where they have failed

[30] George Kistiakowsky, "Science and Foreign Affairs," *Bull. Atom. Sci.*, Vol. 16, No. 4, April 1960, p. 115.
[31] C. W. Sherwin, "Securing Peace Through Military Technology," *Bull. Atom. Sci.*, Vol. 12, No. 5, May 1956, p. 159.
[32] *ibid.*, p. 160.

with theirs. Although this invidious comparison between the scientist's successes and the diplomat's failures seldom is made explicit, the following comment by Harlow Shapley indicates that the feeling is there and that every once in a while it comes through to the surface: "All goes smoothly. We in IGY[33] co-operate; in the UN they expostulate. The musicians are doing a little also, and the cardiologists; student exchanges should increase . . . I pause here a second to exclaim, uselessly, 'curses on the diplomats.' Useless—but it gives me a bit of relief."[34]

As was mentioned above, the physical scientist tends to attribute the difference in relative success between himself and the statesman to the difference in methods and habits of mind rather than to a difference in subject matter. The scientist tends to believe that he, in contrast to the statesman, has learned to rid himself of those intellectual impediments which prevent the former from reaching out for "bold and imaginative" solutions to political problems such as those posed by the development of atomic energy.

Scientists, in fact, have become the third group since the breakdown of the unifying force of medieval Christianity which has sought to bring order to international relations by some non-political means. In the seventeenth century the lawyers such as Grotius, Suarez, and Pufendorf believed international law could regulate the conflicts among nations. In the nineteenth century it was believed that the commercial spirit which could not coexist with war was coming to control the affairs of every nation. Today, depending upon the scientist's specific political view, there is a strong tendency among scientists to believe that the methods, the professional community, and the technical products of science have provided mankind with the means by which to achieve eternal peace.

In the eyes of the student of politics these political attitudes of the scientist possess an ambivalent quality. On the one

[33] IGY is the abbreviation for International Geophysical Year under whose auspices scientists from all nations cooperated in such programs as space, oceanographic, and Antarctic research.

[34] Quoted in Douglas Hurd, "A Case for the Diplomats," *Bull. Atom. Sci.,* Vol. 16, No. 2, February 1960, p. 52.

hand, the scientist contributes to political life, in addition to his intelligence and skills, an optimism and conviction that man's most difficult political problems can be solved; society is deeply indebted to the scientist for the energy he has expended in searching out a meaningful and effective policy toward nuclear weapons. Without this selfless commitment of the scientist to the commonweal, society could not hope to succeed in its struggle to live with atomic energy.

Yet these qualities of concern, persistence, and commitment which characterize the scientist as a political animal have their negative aspects as well. They have meant that rather than having reviewed their premises and proposals analytically, the scientists frequently have tended to reason from necessity. Each deterioration of the prospects for the achievement of *the* solution they propose to the problem of atomic weapons only enhances their determination that this solution *must* and, therefore, *can* be achieved. As a result the scientist in politics often develops a self-reinforcing system of thought which rules out the elementary canons of evidence that he would normally apply in his own field of study. For the scientist the failure to achieve a political goal tends to represent a temporary defeat in his battle against unreason rather than to stimulate a rethinking of his fundamental political position. The consistency of the scientist's political positions over the years and the constancy of his attempts to implement them attest to this generalization.

The Scientists' Conflicting Political Attitudes

Thus far we have considered two types of non-technical assumptions underlying the scientist's advice on national policy toward nuclear weapons which all politically active scientists tend to share. The discussion now turns to a consideration of the fact that different groupings of scientists have conflicting assumptions concerning the most immediate goals to be pursued and the means to be used in the exercise of their social responsibility which have given rise to an intra-scientific conflict over American nuclear weapons policy.

Until about 1947, American scientists were in general agreement upon the Baruch Plan as the solution to the problem of nuclear weapons. Then the failure of the Soviet Union to accept the Baruch Plan and the commencement of the Cold War shattered this scientific unity. The scientific community was further fragmented in 1949 over the issue of the hydrogen bomb. Since that time a fairly constant struggle to influence American nuclear policy in divergent directions has taken place among three groupings of scientists. The conflict among scientists over the issues of the hydrogen bomb, tactical nuclear weapons, and a nuclear test ban have all been manifestations of this important struggle.

During these latter disputes the American scientific community, or at least the politically active segment of it, has been fragmented into three schools of thought: the control school represented by Linus Pauling; the finite containment school represented by Hans Bethe; and the infinite containment school represented by Edward Teller. The basic causes of the political cleavage over nuclear weapons policy among these schools of scientists have been their essentially opposed assumptions on the nature of the nuclear arms race, the nature of Russian intentions, and the nature of international politics.[35] On the basis of their respective views the different schools of scientists have at times come to conflicting positions on the wisdom of alternative courses for American nuclear policy. Of necessity these views have influenced the advice given by scientists on such matters as the hydrogen bomb and a nuclear test ban.

It should be noted that the term "school" is used to set apart a grouping of men who share a common intellectual perspective or set of assumptions. It does not deny the in-

---

[35] Whether these divisions among the scientists on nuclear weapons policy reflect underlying socio-economic sources of cleavage is a subject beyond the scope of this study.

dividuality of the scientists nor the differences which do exist amongst members within each grouping; each school of thought instead denotes those who share a common disposition toward the issue of American policy toward nuclear weapons, a disposition which is distinct from that of other scientists. Furthermore, the terminology "school of scientists" as used herein does not mean a cabal or—less dramatically—an interacting group of men, even though the scientists in one school or another have upon occasion cooperated for political purposes. The important point is that "school" indicates basically an intellectual position and only on occasion a group of scientists actively cooperating in pursuit of a common goal.

The persistence of these schools of thought and the regularity with which scientists fall into one or another appear to result from two factors. They reflect the scientists' strong emotional conviction of the need to solve the problem of atomic weapons and the individual scientist's lack of time or aptitude to pursue the subject independently. Therefore, in place of suspended judgment or of independent analysis, the scientists in general accept the views of the few scientists such as Hans Bethe, Linus Pauling, and Edward Teller who have given extensive thought to the issue of nuclear weapons. Consequently, as was mentioned earlier, when the ideas of an individual scientist are presented in this study, it should be kept in mind that they generally represent a significant body of scientific opinion.[36]

Political Leadership and the Dilemma in American Nuclear Policy

The intensity of the intra-scientific conflict over nuclear weapons is symptomatic, if not the result, of a serious failure of American political leadership with respect to nuclear

---

[36] There are, of course, individual scientists who do not fit into any of the three groups of scientists focused upon in this study. Vannevar Bush and Herman Kahn are of this stripe. See Vannevar Bush, *Modern Arms and Free Men*, Simon and Schuster, 1949, and Herman Kahn, *On Thermonuclear War*, Princeton University Press, 1960.

weapons. Since 1947 American nuclear policy has been characterized by a dangerous contradiction. At the same time that the goal of American disarmament policy has been the elimination of atomic weapons, America's military policy has been over-dependent upon atomic weapons for the defense of its foreign policy commitments. This situation brought about a dilemma in American nuclear policy from which the nation is only slowly extracting itself.

The United States could afford such a discrepancy between its disarmament policy and its military policy when there was little likelihood that one would interfere with the other. For example, the Baruch Plan was no threat to the implementation of the recommendation of the Commission on Air Power[37] that American defense be based on the strategic employment of atomic bombs, because in the years immediately following World War II there was little likelihood that the Soviet Union would subscribe to the Baruch Plan. However, as disarmament or arms control plans with respect to atomic weapons such as a nuclear test ban became more definite and subject to implementation while the military threat from the Soviet Union increased, the contradiction in these American policies could no longer be ignored.

American scientists have often felt impelled to provide the leadership in attempts to resolve this dilemma. Given the indecisiveness and technical unawareness of American political leaders with respect to nuclear weapons, the views of scientists have had a significant influence. As will be seen, all but one of the major departures in American policy toward nuclear weapons were initially conceived by scientists: the Baruch Plan, the hydrogen bomb, the development of tactical nuclear weapons, the ballistic missile, and a nuclear test ban. Only the doctrine of massive retaliation originated elsewhere.

Unfortunately, American political leadership has not always recognized aspects of a political nature in the scientist's

[37] U.S. President's Commission on Air Power, *Survival in the Air Age*, U.S.G.P.O., 1948.

advice. Viewing the scientist solely as a technical person, political leaders have at times uncritically accepted the advice of the scientist when it has agreed with their current political predilections. Yet, when the advice of the scientist has found disfavor, it has frequently been rejected without careful evaluation regardless of its merits.

The failure to maximize the potential benefits of the scientist's entry into political life has been both the cause and the effect of the lack of a coherent and realistic American policy toward nuclear weapons. If the United States is to fashion an effective policy toward nuclear weapons and other products of the scientific revolution the scientist has to be utilized in a far more realistic manner than has been the case thus far. Toward this end the present study of the intra-scientific conflict over nuclear weapons and its effect on American nuclear policy is directed.

CHAPTER TWO

SCIENTISTS SEARCH FOR A
BETTER WORLD

The Genesis of the Scientists' Views
on Nuclear Weapons

ALTHOUGH a considerable number of scientists did not become concerned about the political, military, and moral implications of the development of atomic energy until after the explosion of the first atomic bomb, there had been significant groups of outstanding scientists who, during their years of wartime research, had devoted a great deal of thought to the problems which their discoveries portended for the future.

The ideas generated by these scientists were to become the basis for a temporary unity among politically active scientists in their early postwar effort to obtain international control of atomic energy; these same ideas are still influential today in determining the attitudes of scientists toward nuclear weapons and have become important ingredients in American nuclear policy.

THE UNDERLYING CONCERN OF THE SCIENTISTS

The scientists who participated in wartime discussions of novel problems of the atomic age were motivated by the dire consequences they foresaw as possible results of their development of atomic weapons. They were acutely aware that if some system of international control over atomic energy were not established soon after the atomic bomb was revealed to the world, the military services of all nations would seek to achieve such weapons and thus give rise to a potentially disastrous nuclear arms race.

As early as 1943 a number of these scientists had realized that the atomic bomb was an innovation which would be the ideal weapon of total war, of surprise attack, and of aggression. The atomic bomb would surely be recognized as the ideal weapon for the type of military operations envisaged by the doctrine of strategic air power, a doctrine which reduced all military operations to the air strike against the enemy nation and which was beginning to be put into operation by the Allies.

Although the doctrine of strategic air power had its origins in World War I and was acted upon by the Army Air Corps in World War II, it did not become pre-eminent until after 1945. In fact, on the basis of World War II experience, strategic bombing was so ineffective in comparison to the effort expended that its future was actually in doubt until the explosion of the atomic bomb on August 6, 1945. At that time the doctrine of strategic air power was quite literally rescued by the result of the Manhattan District Project, and soon, much to the scientists' concern, this doctrine was enthroned as the dominant component in Western military strategy.[1]

Many of those scientists who participated in the Manhattan District Project believed that once the atomic bomb were proved feasible, there would be no way to prevent other nations from developing and utilizing atomic weapons. Even in a divided and secretive world there was no secret to the atomic bomb which the United States could hope to withhold from other nations for long. The secrets of nature are accessible to competent scientists in all nations. The only conceivable secret was that an atomic bomb was possible; American use of the bomb would demonstrate this to the whole world.

Even the lead of the United States in the atomic weapons field, these scientists felt, would be no guarantee of its security in the long run. There would be no advantage to the United States in possessing a superior number of atomic weapons;

[1] For a thorough history and critique of the doctrine of strategic air power see Bernard Brodie, *Strategy in the Missile Age,* Princeton University Press, 1959.

once the devastating weapon had been developed by other nations it would enable even a small aggressive nation to destroy the industrial capability of the United States.

The only defense against atomic weapons that these scientists could foresee was the wide dispersal of population and industry. Thus, although all nations would be vulnerable to atomic attack, the United States appeared to be especially vulnerable because of its high urban concentration. The United States, the arsenal of democracy, would be put at a great disadvantage by the weapon it was about to introduce into the world. Democracy's great resource was America's vast industrial capacity which had enabled it and its allies to win two world wars. This productive capacity had always been secure behind two oceans and the United States had had sufficient time available in which to produce the weapons of war and to meet aggression.

The atomic bomb would destroy this security based on productive capacity and geographical isolation. The atomic bomb and the airplane together meant that the highly concentrated industrial capacity of the United States could be wiped out by an aggressor in a surprise attack. Moreover, the atomic bomb, because it is indeed the ideal weapon of surprise attack and of the offense, seemed to be disadvantageous to a deliberative and non-aggressive democracy such as the United States. Furthermore, the scientists believed that fear of retaliation by the United States would not necessarily prohibit surprise attack because such an attack might be able to destroy America's ability to retaliate.

In such a world the scientists believed that the insecurities of an atomic arms race would slowly erode the democratic process. The military would begin to control everything, including science, in its desire to create superior weapons. Eventually, they feared, this would result in accidental or premeditated nuclear war which would destroy civilization.

The foregoing ideas and proposals had been thoroughly discussed by the scientists in the Manhattan District Project prior to the explosion of the atomic bomb over Hiroshima. Embodied in what have become known as the Bohr Memorandum and the Franck Report, they would influence the positions of scientists on all subsequent issues involving atomic weapons. While these ideas were to provide a source of unity within the scientific community with respect to nuclear weapons in the early post-war period, differences merely in the emphasis within the two major reports would become sources of deep cleavage with the advent of the Cold War and the nuclear arms race.

THE BOHR MEMORANDUM

The foremost of the scientists as well as one of the first to give profound attention to the implications of atomic energy was the "father of atomic physics," Niels Bohr. In the theoretical division at Los Alamos, along with his associates who included his younger former colleagues Robert Oppenheimer and Edward Teller, Niels Bohr explored the frightening possibilities of the coming atomic age with its military weapons of greatly increased destructive power. And in 1944 Bohr put his thoughts down in a memorandum which he presented to President Franklin Roosevelt.[2]

Although Bohr's statement had little effect on the wartime President, it did have and continues to have a strong influence on his scientific colleagues. Bohr prophesied that atomic energy would not only "revolutionize industry and transport" but that it would "completely change all future conditions of warfare;" furthermore, the terrifying prospects for future warfare could only be dimly perceived because, immense as it was, the wartime research effort was only the beginning and research "continually revealed new possibilities."

Bohr specifically warned the President that any temporary advantage which the United States would gain through possession of the atomic weapon would be "outweighed by a

---

[2] As it appeared in Robert Jungk's *Brighter Than a Thousand Suns*, Harcourt, Brace and Company, 1958, pp. 344–47.

perpetual menace to human security." While he hopefully envisioned the possibility of a "world-wide scientific collaboration which for years has embodied such bright promises for common human striving," Bohr believed that the frighteningly destructive possibilities of the atomic age were very real and could be avoided only with very careful planning and constant effort by the nations.

Bohr then foreshadowed the crux of the problem of nuclear weapons as it would be viewed by all politically active scientists in the postwar period: "The prevention of a competition prepared in secrecy will therefore demand such concessions regarding exchange of information and openness about industrial efforts, including military preparations, as would hardly be conceivable unless all partners were assured of a compensating guarantee of common security against dangers of unprecedented acuteness."[3]

Bohr was suggesting that the only true security in the postwar atomic age would be an open world, free from secrecy, where atomic energy was under some form of international control. He argued that in an age where science had already harnessed atomic energy and would continue to create new means of military power, openness and atomic energy control were the only guarantees of security and a lasting peace.

In Bohr's view, the world was only in the first phase of the scientific revolution. As man's knowledge of the physical world increased anything might be possible. In such a world, Bohr argued, there could be no true security unless each nation were fully informed of the scientific and technical developments taking place in all other nations. Just as scientific advance had made possible the secret development of the atomic bomb, other devastating weapons might become possible as man's knowledge increased. Each nation, to be secure, would have to be sure other nations were not secretly exploiting scientific advances for aggressive purposes.

While he disqualified himself from possession of a com-

[3] *ibid.*, p. 345.

petence to solve the political problems involved, Bohr was optimistic about the possibility that the nations could be induced to accept scientific openness and international control over atomic energy. His confidence rested on the expectation that nations would realize that they had more to gain from the peaceful exploitation of science including peaceful applications of atomic energy than they had to gain from engaging in an ominous nuclear arms competition.

THE FRANCK REPORT

Scientists at the other Manhattan District Project sites also discussed the consequences for the world of the harnessing of atomic energy. At the Metallurgical Laboratory of the University of Chicago this thinking reached a rather systematic formulation. Scientists there had completed their work assignment by late 1944 and had time available to think about the impending results of their labors. The principal grouping of these scientists, known as the Committee on Social and Political Implications, was headed by James Franck and included such men as Glenn Seaborg, Leo Szilard, C. J. Nickson, and Eugene Rabinowitch.

The Metallurgical Laboratory scientists sent a number of petitions to the President and his advisors proposing policy toward atomic weapons, the most important of which was the Franck Report.[4] Although the Report paralleled much of Bohr's memorandum, it differed from this earlier document in a number of significant respects. For example, while the scientists who wrote the Franck Report also disclaimed any presumption "to speak authoritatively on problems of national and international policy," in contrast to Bohr, they felt a responsibility "to offer . . . some suggestions as to the possible solution of these grave problems." These scientists based their competence to make policy suggestions upon their "acquaintance with the scientific elements of the situation and

[4] For a copy, see *ibid.*, pp. 348–60.

prolonged preoccupation with its world-wide political implications . . . ."[5]

Specifically, the Franck Report argued that the primary
political-military aim of the United States in the postwar
world should be the prevention of an atomic arms race and
that the international control of atomic energy provided the
only means through which an atomic arms race could be
prevented. These scientists felt that if all other nations were
provided with the facts about atomic weapons they too would
inevitably reach this conclusion. To educate the United
States and other nations to these facts was their responsibility
as scientists.

While the specific nature of the control system which
these scientists proposed is not of interest here, these men
did believe that "one thing is clear: any international agreement on prevention of nuclear armaments must be backed by
actual and efficient controls. No paper agreement can be
sufficient since neither this nor any other nation can stake its
whole existence on trust in other nations' signatures."[6] It will
be seen that this conviction is basic to the thinking of most
scientists as well as the United States Government in the
postwar disarmament negotiations.

Following the discussion of the desirability of achieving
international control, the report considered the questions of
the technical feasibility of control, of means for achieving it,
and of alternative proposals. The report at this point provides
significant insights into ideas which continue to influence the
political activities of a number of prominent scientists.

The Franck Report reasoned that because every nation,
including the Soviet Union, had an interest in self-preservation,
they would all "shudder at the possibility of a sudden disintegration . . . ." The report then reasoned: "Therefore,
only lack of mutual trust, and not lack of desire for agree-

[5] *ibid.*, p. 349.
[6] *ibid.*, p. 358.

ment, can stand in the path of an efficient agreement for the prevention of nuclear warfare. The achievement of such an agreement will thus essentially depend on the integrity of intentions and readiness to sacrifice the necessary fraction of one's own sovereignty, by all the parties to the agreement."[7]

These scientists recommended that the obvious truth of this generalization should dictate all American policy on nuclear weapons, including the determination of whether or not the atomic bomb should be used to hasten the end of the war. The authors of the Franck Report believed that the United States was not committed to use of the bomb when completed, but had constructed the atomic bomb solely to deter its use by any other nation. "The compelling reason for creating this weapon with such speed was our fear that Germany had the technical skill necessary to develop such a weapon and that the German Government had no moral restraints regarding its use."[8]

Although these scientists ruled out the advisability of the use of the atomic bomb against Japan, they did suggest two alternatives to their political superiors. One was that, if the prospects for international control were judged to be favorable, the United States ought just to demonstrate to the world and especially to the Japanese, the power of the atomic bomb. This would bring the war to an end without prejudicing the accomplishment of international control. If, on the other hand, the government believed that there was little chance of achieving international control, a public demonstration that the atomic bomb was technically feasible would mean "a flying start towards an unlimited [nuclear] armaments race." With this possibility in mind the scientists recommended that the government should consider the alternative that: "The benefit to the nation, and the saving of American lives in the future, achieved by renouncing an early demonstration of nuclear bombs and letting the other

[7] *ibid.*, p. 354.
[8] *ibid.*, p. 357.

nations come into the race only reluctantly, on the basis of guesswork and without definite knowledge that the 'thing does work,' may far outweigh the advantages to be gained by the immediate use of the first and comparatively inefficient bombs in the war against Japan."[9] Thus the United States would renounce a temporary military advantage in favor of preventing a postwar nuclear arms race.

The Call to Action: Hiroshima and the May-Johnson Bill

THE DECISION TO DROP THE BOMB

While the project scientists had been holding their discussions other persons were considering related questions. Shortly after he became President, Harry Truman, in anticipation of the success of the Manhattan District Project, had appointed an Interim Committee under the chairmanship of Secretary of War Henry Stimson to advise him on two questions. The first question dealt with the postwar disposition of atomic energy; the second question was whether or not the United States should use the atomic bomb against Japan.

The Interim Committee which had to make specific policy recommendations included among others the leaders of the wartime scientific effort, namely Vannevar Bush, Karl Compton, and James Conant. To assist it, the Interim Committee appointed a panel of scientific advisors composed of Robert Oppenheimer, Arthur Compton, Ernest Lawrence, and Enrico Fermi. All of these scientists except Fermi had played leading administrative roles throughout the war; they all had borne a major responsibility for the development of the atomic bomb and were now to share in the decision that it should be used against the Japanese.

After a long period of deliberation during which the Interim Committee considered and rejected a variety of policies, it was finally decided that the only realistic course to

[9] *ibid.*, p. 356.

follow was full military use of the bomb. As regrettable as such an action would be, it seemed evident to these men that it was the only justifiable decision. The questionable impact which a peaceful demonstration of the atomic bomb might have on Japanese opinion, the apparently ferocious determination of the Japanese to fight to the bitter end, and the blood bath which an invasion of Japan would cause were some of the considerations which led to this fateful decision.[10]

Essentially, the view of the Interim Committee and its scientific advisory committee was that the atomic bomb was a proper, even though a very powerful, weapon. Contrary to the position of the Franck Report that the United States had developed the atomic bomb only to deter its usage by the Germans, Stimson and his colleagues did not question, at least for long, the military use of the bomb. "The possible atomic weapon was considered to be a new and tremendously powerful explosive, as legitimate as any other of the deadly explosive weapons of modern war."[11]

The position of the scientists who participated in this decision to drop the bomb was summarized in a formal report to Stimson from the scientific advisory committee: "The opinions of our scientific colleagues on the initial use of these weapons are not unanimous; they range from the proposal of a purely technical demonstration to that of military application best designed to induce surrender. Those who advocate a purely technical demonstration would wish to outlaw the use of atomic weapons, and have feared that if we use the weapons now our position in future negotiations would be prejudiced. Others emphasize the opportunity of saving American lives by immediate military use, and believe that such use will improve the international prospects, in that they are more concerned with the prevention of war than with

---

[10] A number of writers have subjected this decision to scrutiny. For an excellent concise treatment, see Herbert Feis, *Japan Subdued: The Atomic Bomb and the End of the War in the Pacific,* Princeton University Press, 1961, Chapter Four.

[11] Henry Stimson, "The Decision to Use the Bomb," *Harper's Magazine,* Vol. 194, No. 1161, February 1947, p. 98.

the elimination of this special weapon. *We find ourselves closer to these latter views; we can propose no technical demonstration likely to bring an end to the war; we can see no acceptable alternative to direct military use.*"[12]

Thus arose the first political cleavage among the scientists over nuclear weapons policy. The authors of the Franck Report, speaking for many Project scientists, and the scientists on the Interim Committee and its scientific advisory committee differed on two fundamental issues: (1) Should the United States drop the bomb on Japan? (2) Should the goal of the emerging American nuclear policy be to eliminate nuclear weapons or to utilize the threat of nuclear weapons to eliminate war? While the first issue was soon to become a matter of history, the second issue would reappear to divide the scientists following the outbreak of the Cold War and it continues to do so to the present day.

THE REACTION OF PROJECT SCIENTISTS TO HIROSHIMA

Throughout the war, scientists in administrative positions like those who sat on the Interim Committee and its scientific advisory panel had in effect spoken for science. The rank and file, because of their subordinate position and the requirements of secrecy, had had little influence on the utilization of science in the war effort. As a result many believed, somewhat erroneously, that their views had either been disregarded or distorted by the scientific leadership. Furthermore, it seemed to many of the rank and file scientists that the vital consequences of the military revolution wrought by atomic energy were beyond the understanding of their elders who were advising the government. The explosion of the atomic bomb convinced them of this at the same time that it released them from the bondage of secrecy.

These feelings of discontent were translated into political action when the rank and file scientists learned that certain persons in the Administration and in the Congress were trying

[12] Herbert Feis, *op.cit.,* p. 43. (Italics supplied by Henry Stimson.)

to rush through Congress a bill which purportedly would give the military continued control over atomic energy. The ire of the Project scientists was further aroused when they discovered that the wartime leadership of science supported the proposed May-Johnson Bill which would allegedly "militarize" atomic energy. From Oak Ridge, Los Alamos, and the Metallurgical Laboratory (Chicago), scientists converged upon the nation's capital to lobby against the May-Johnson Bill. This bill which had been proposed by the Interim Committee would have established an Atomic Energy Commission with rather arbitrary powers over the field of atomic energy, including scientific research. The reaction of most scientists was one of fear as they were convinced that military officers would come to administer science and would impose serious restrictions on dissemination of scientific knowledge.[13]

The movement of the rank and file scientists quickly became one to convince the Administration, the Congress, and the nation of the need to achieve the international control of atomic energy which had been outlined in the Franck Report and the Bohr Memorandum. In order to achieve this political objective these scientists formed various organizations such as the Emergency Committee of Atomic Scientists, the Federation of Atomic Scientists, and the National Committee on Atomic Education. These groups were not divided on policy objectives but rather represented a division of labor in the efforts of the scientists to attain international control; one concentrated on fund raising, another on political action, and the third on education. Often there were interlocking directorates among these groups and always the dominant leadership was provided by Project scientists, especially the physicists.

---

[13] The history of this period is exceedingly complex and will certainly not be untangled until the AEC completes its official history on the struggle over the establishment of the agency. A valuable short history of this period and many other episodes covered by the present study is J. Stefan Dupré and Sanford Lakoff, *Science and the Nation—Policy and Politics,* Prentice Hall, 1962.

Essentially the scientists saw the problem of the control of atomic energy as a technical one, and they believed that they had a responsibility to educate mankind in general and the politicians in particular to what they conceived to be the only technical answer to this problem, the establishment of international control. Despite their deep involvement in politics, however, these scientists denied quite vociferously that they were advocates of a decidedly political position. During the height of the struggle to convert the United States Senate to the principle of international control the President of the Federation of Atomic Scientists stated: "The Federation makes a point of being non-political. . . . To hell with politics. The question is: Are you pro- or anti-suicide?"[14]

The greater part of the energies of these scientists was directed toward the establishment of an atomic energy commission which would meet their requirements. Specifically they desired a civilian controlled atomic energy commission which would symbolize to the rest of the world the desire of the United States to avoid an atomic arms race and would actually be a first step toward the international control of atomic energy. A United States Government monopoly over the American atomic energy enterprise could simply be transferred to an international agency following an appropriate agreement among nations.

The scientists believed that the clock of man's history stood at a "few minutes to midnight."[15] This was a gripping belief which the scientists were able to communicate to the public, the Congress, and the Executive through hearings, lobbying, and speeches. The scientists' conviction of the rightness of their cause, together with their newly gained prestige impressed Washington greatly. The men "who built the bomb" did much to convince the nation that it must bring atomic energy under some form of international control.

[14] Daniel Lang, "That's Four Times 10^{-4} Ergs, Old Man," *The New Yorker*, November 16, 1946, p. 78.

[15] This symbol, suggested by Edward Teller, appeared as the famous clock on the front of the *Bulletin of the Atomic Scientists*.

In retrospect, the truly amazing thing is not that these scientists were not consulted more extensively prior to the decision to drop the bomb, but that they and their colleagues on the Interim Committee and its scientific advisory panel were consulted at all. It was most unusual that scientists were called upon and were listened to on questions of the correctness of a political-military decision. This occurrence presaged the forthcoming revolution in the relationship of the scientist to national decision-making.

The Attempt to Achieve International Control of Atomic Energy: The Baruch Plan

THE POSITION OF THE UNITED STATES

A number of individuals within the Truman Administration had also been giving increasing attention to the problem of postwar policy toward nuclear weapons. The most prominent and significant of these persons were Henry Stimson, the Secretary of War, and Vannevar Bush, the Director of the Office of Scientific Research and Development. These men, presumably influenced by the thinking of such scientists as Niels Bohr and Leo Szilard, supported the notion of the international control of atomic energy.

Through the influence of Stimson, Bush, and others as well, the indecisiveness which had originally characterized American postwar nuclear policy was eventually resolved in favor of an attempt to bring atomic energy under some system of international control. This victory was symbolized by the Truman-Attlee-King Declaration of November 15, 1945, which had been drawn up by Vannevar Bush and which by implication committed the United States to the international control of atomic weapons.[16] This declaration called upon the United Nations Organization to establish a commission with the task of fashioning proposals to bring about an open

[16] U.S. Atomic Energy Commission, *In the Matter of J. Robert Oppenheimer*, U.S.G.P.O., 1954, p. 561.

world, to promote the peaceful uses of atomic energy, and to eliminate nuclear weapons from the world.

Furthermore, President Truman assigned to Under Secretary of State Dean Acheson the responsibility of developing a still more detailed policy toward nuclear weapons. Acheson in turn appointed a high-level advisory panel under David Lilienthal to formulate the United States position on atomic energy. Among those selected to serve with Lilienthal in the formulation of the American position was the wartime scientific director at Los Alamos, Robert Oppenheimer. Through Oppenheimer's participation in the panel many of the ideas generated by the scientists during the war were fashioned into American policy.

The report of the Lilienthal Committee, which was to be known as the Acheson-Lilienthal Proposals, became the basis for the United States plan for the international control of atomic energy. This plan, which eventually took its name from Bernard Baruch, the senior United States representative in the United Nations atomic energy negotiations, sought to achieve a limited world government in the area of atomic energy.[17] It proposed an Atomic Development Authority which would have a monopoly of all the world's "dangerous" fissionable materials and production plants. Any attempt by a nation to build atomic weapons would be known to the Authority which would report such a violation to the nations of the world so that appropriate sanctions could then be applied against the offender. Furthermore, the Atomic Development Authority would have the right to develop the peaceful uses of atomic energy; it would be a positive force for

[17] This study makes no distinction between the Acheson-Lilienthal Proposals and the Baruch Plan despite the fact that the plan provided for sanctions against violations. The provision for sanctions was not crucial to the outcome of the UN debate; also the scientists who are our main concern supported the Baruch Plan vehemently against critics of the sanctions provisions such as Vice President Henry Wallace. For an excellent critique of the Baruch Plan see Robert Gard, "Arms Control Policy Formulation and Negotiation, 1945–1946" (Ph.D. thesis, Harvard University, 1961).

peace through helping to remove economic want as a cause
of war.

Reduced to its essence the Baruch Plan was a mechanism
to prevent surprise attack, to maintain an "open world" in
nuclear research, and to preserve the international *status quo*
by freezing military technology at its World War II level.
Although the proposed Authority would not actually have the
power itself to prevent a nation from converting atomic
energy into military weapons, at least it would guarantee
that such conversion could not be done in secret. In effect,
then, the violation of the treaty would serve notice that a
nation was preparing for nuclear war. As the Acheson-Lilienthal Report stated, "The security which we see in the realization of this plan lies in the fact that it averts the danger of
the surprise use of atomic weapons."[18]

Significantly, the Baruch Plan also provided for an open
scientific world in that all the research laboratories under the
Authority, wherever they were located, would be open to the
scientists of all nations and scientists in nuclear physics would
be free to communicate with other scientists. The political
significance of such freedom of communication would be
that nations would be prevented from taking secret advantage
of new knowledge. Scientific breakthroughs which would
enable a nation to infringe on the control system established
under this plan would be known to all and the control system
could be improved as fast as knowledge developed. Every
nation would thus be assured that no other was secretly advancing its nuclear weapons technology.[19]

The importance of an open world to the success of the
international control of atomic energy was explained by
Oppenheimer in 1948: "Let me mention 1 or 2 points. One,
to my mind, the principal one, was that it was clear that
no secure system could be developed for protecting people

[18] U.S. Department of State, *A Report on the International Control of
Atomic Energy,* U.S.G.P.O., 1946, p. 53.
[19] See Oppenheimer's discussion of this point in U.S. Atomic Energy
Commission, *In the Matter of J. Robert Oppenheimer, op.cit.,* p. 37.

against the abuse of atomic weapons, unless the world were open to access, unless it was possible to find out the relevant facts everywhere in the world which had to do with the security of the rest of the world. This notion of openness, of an open world, is, of course, relevant to other aspects of the Soviet system. It is doubtful whether, without the newly terrible, yet archaic, apparatus of the Iron Curtain, a government like the Soviet Government could exist. It is doubtful whether the abuses of that Government could persist."[20]

The Baruch Plan would have preserved the traditional American military advantages of its military-industrial capacity and geographical isolation which an atomically armed world would cancel. With this plan in force, the United States could have remained secure behind its oceanic barrier, assured that no nation could destroy American military-industrial power by surprise attack, and it could have continued to enjoy freedom from the military insecurity and large standing armies which had plagued the Old World. The price the nation would have had to pay for this guarantee would have been the relinquishment of its atomic monopoly and participation in an international program to develop the peaceful uses of atomic energy.

This interpretation of the Baruch Plan as a mechanism to prevent surprise attack and to maintain the *status quo* of the United States as the "arsenal of democracy" is supported by the Acheson-Lilienthal Report itself: "Our political institutions, and the historically established reluctance of the United States to take the initiative in aggressive warfare, both would seem to put us at a disadvantage with regard to surprise use of atomic weapons. This suggests that although our present position, in which we have a monopoly of these weapons, may appear strong, this advantage will disappear and the situation may be reversed in a world in which atomic armament is general."[21]

[20] *ibid.*, p. 45.
[21] U.S. Department of State, *A Report on the International Control of Atomic Energy, op.cit.*, pp. 2–3.

The Atomic Development Authority proposed by the Baruch Plan would, in reality, have constituted a "trip wire." Although it could not have stopped a nation from seizing and utilizing for aggressive purposes the nuclear materials within its own borders, the Authority would have prevented any nation from doing this without providing the other nations with adequate warning time in which to rebuild their nuclear forces. Thus, while the Plan could not have prevented war, it could have insured a lag-time of approximately one year between a nation's breach of the treaty and its attainment of a capability to wage atomic warfare; this, in the words of the Acheson-Lilienthal Report, would have given the United States the time to "rebuild its forces and prepare for the attack."

If the Baruch Plan were to achieve its end of freeing the world and especially the United States from the threat of surprise attack, it was essential that there be a carefully coordinated phasing of the release of technological information on atomic energy with the imposition of international political controls on the use of such information. As was pointed out in the Report: "The significant fact is that at all times during the transition period at least such facilities will continue to be located within the United States. Thus should there be a breakdown in the plan at any time during the transition, we shall be in a favorable position with regard to atomic weapons."[22]

Thus, despite its daring and idealistic enlightenment, the Baruch Plan was truly a reflection of American national interest. The United States had nothing to lose and everything to gain through implementation of the Baruch Plan in a world of rapid technological advance where America's monopoly on atomic energy was at best a transient one. In return for long-run security, the United States would merely

[22] *ibid.*, p. 57.

share sooner the atomic secrets upon which its power position was temporarily based.

THE POSITION OF THE AMERICAN SCIENTISTS

The accomplishment of the Lilienthal Committee was to integrate the views of those scientists (such as the signers of the Franck Report) who favored the total elimination of nuclear weapons, with those of other scientists (like the members of the scientific advisory panel of the Interim Committee) who believed the threat of nuclear weapons would force man to eliminate war. While the overt purpose of the Baruch Plan was to prevent a nuclear arms race, the scientists reasoned hopefully that mutual fear of nuclear weapons would prevent all war lest it become nuclear. For this reason the Acheson-Lilienthal Proposals and the Baruch Plan unified the scientific community in the performance of its social responsibility with respect to nuclear weapons during the early postwar years.

The enthusiastic support of most American scientists for the Baruch Plan was not, then, simply a chauvinistic acceptance of the American position but an outgrowth of the deep conviction that the American plan was commended by the voices both of reason and of science. In contrast to all "political schemes," the Acheson-Lilienthal Proposals upon which the Baruch Plan was based were accepted as the result of the application of the scientific method to politics. It was believed that the panel which had drawn up the Plan had discarded all preconceptions and followed only the "facts" of atomic energy.

The faith of scientists in the irresistible nature of their arguments for the Baruch Plan was expressed graphically by Eugene Rabinowitch in the *Bulletin of the Atomic Scientists:* " . . . when they [non-scientists] are called to devise a program, they at first are bound to shy away from propositions which appear too radical and unrealistic, and to try to find solutions of a more conventional character. However, we have seen during the past nine months that the inescapable *'logic of facts'* ends in converting most if not all who have studied the problems, from Sauls into Pauls of international control."[23]

The convergence of the official plan of the United States with the ideas expressed in the Franck Report, the Bohr Memorandum, and the report of the Interim Committee advisors reinforced the belief of the scientists that the facts of atomic energy compelled acceptance of the Baruch Plan by all rational men. They were certain that, just as any scientific fact or theory compels its own acceptance over time, so would the United States proposals eventually be accepted by all.

The confidence of the scientists and others in the ultimate success of the Baruch Plan rested upon their belief that it was the result of the application of the scientific method to politics. Rather than following the allegedly traditional political method of starting with a "preconceived 'solution'," the Acheson-Lilienthal Committee had pursued "a patient and time-consuming analysis and understanding of the facts;"[24] therefore, the proposals were viewed as the logical consequence of these facts. His committee, in the words of Lilienthal, had had "an opportunity to analyze what is called a political problem in a scientific spirit . . . we started somewhat as a chemist might, tackling a technical problem: with the facts as he found them."[25] It was believed that other men in other nations would come to the same conclusion on this problem if they employed the same technique.

The Acheson-Lilienthal Committee believed that the ra-

[23] "A Dangerous Lull," *Bull. Atom. Sci.,* Vol. 1, No. 11, May 15, 1946, p. 1. (Italics mine.) Rabinowitch then goes on to recount how this conversion took place for the persons who drew up the Acheson-Lilienthal Proposals.

[24] U.S. Department of State, *A Report on the International Control of Atomic Energy, op.cit.,* p. XI.

[25] David Lilienthal, "How Can Atomic Energy Be Controlled?" *Bull. Atom. Sci.,* Vol. 2, Nos. 7 & 8, October 1, 1946, p. 15.

tional method they had employed was a contribution to international politics as important as the substance of their actual conclusions. Their report was intended to open the eyes of the nations to the value of this method: "Five men of widely differing background and experiences who were far apart at the outset found themselves, at the end of a month's absorption in this problem not only in complete agreement that a plan could be devised but also in agreement on the essentials of a plan. We believe others may have a similar experience if a similar process is followed.

"We have described the process whereby we arrived at our recommendation, to make it clear that we did not begin with a preconceived plan. There is this further reason for describing this process. Others would have a similar experience if they were able to go through a period of close study of the alternatives and an absorption in the salient and determining facts. Only then, perhaps, may it be possible to weigh the wisdom of the judgment we have reached, and the possibilities of building upon it."[26]

The initial task which the scientists set for themselves was therefore the education of mankind to the facts of atomic energy. The scientists believed that all men, once they had been made aware of these facts, would understand the wisdom of the Baruch Plan as the solution to the problem of atomic energy. Although the scientists believed that they knew the difficulties to be met in the achievement of their goal, they believed that "the fact that . . . seem[s] to make the problem . . . hopeless, is precisely the central vital fact which makes the solution possible at all."[27] The very immensity and gravity of the problem led the scientists to believe that as

[26] U.S. Department of State, *A Report on the International Control of Atomic Energy, op.cit.,* pp. XI–XII. This notion that technical discussions can be substituted for political negotiations reappears after 1955 and culminates in the 1958 Conference of Experts on the monitoring of a nuclear test ban.
[27] Robert Oppenheimer, "The Atom Bomb as a Great Force for Peace," *The New York Times Magazine,* June 9, 1946, p. 60.

the problem *must* be solved, therefore it *would* be solved. The threat to man's survival would drive men to reason.[28]

If the Baruch Plan should succeed, and the scientists believed it would, not only would the problem of atomic energy be solved, but the example set by the method employed in its solution would profoundly alter the conduct of international relations. In particular, to scientists such as Robert Oppenheimer the Baruch Plan seemed "an opportunity to cause a decisive change in the whole trend of Soviet policy, without which the prospects of an assured peace were indeed rather gloomy, and which might well be, if accomplished, the turning point in the pattern of international relations."[29]

THE POSITION OF THE SOVIET UNION

In retrospect it appears quite evident that the Soviet Union was guided by a different "logic of the facts" from that of the American scientists or government. The Soviet Union, believing time to be on its side, had no incentive to make the great sacrifices of its sovereignty required by the Baruch Plan. The American monopoly of atomic energy was temporary; it would only be a matter of time and effort before the Soviet Union would break that monopoly and possess its own nuclear capabilities.[30]

The Soviet Union, as a revisionist power with respect to geography and to its technological position in the atomic energy field, could not accept a plan which at all times would preserve the United States' technological and geographical advantage while simultaneously opening the Soviet Union to "espionage." Since its plans were dependent upon its offensive success, the USSR could not accept a system which left intact the traditionally strong defensive posture of the United States,

[28] This notion also reappears after 1955 and becomes a strong ingredient in the defense of the feasibility of a nuclear test ban.

[29] Robert Oppenheimer, "International Control of Atomic Energy," *Foreign Affairs,* Vol. 26, No. 2, January 1948, p. 245.

[30] For a more complete analysis, including Soviet fear of capitalist management of the Soviet economy, see Richard Barnet, *Who Wants Disarmament?,* Beacon Press, 1960.

nor could it lay bare the military-industrial capabilities of the USSR to the threat of American strategic air power. The Soviet response to the plan therefore was to attempt to "split the package" and to rephase its implementation to the Soviet advantage.

Thus there were several basic differences in the positions of the United States and the Soviet Union on international controls. The United States wanted the establishment of a control system to precede the prohibition of atomic weapons. Before the Soviet Union would accept any control system it demanded that first the nations agree to the outlawing of atomic weapons. Following such a prohibition, the Soviet Union would then "consider" with the United States a covenant to control atomic weapons. However, by "control" the Soviets meant only a system of mutual "periodic" inspections of nationally owned facilities. In effect, the Soviet Union rejected committing itself to such openness as was envisioned by the system of international control contained in the Baruch Plan.

Instead of agreeing to the need for a decrease in secrecy the Soviet Union actually multiplied its use of secrecy. The Soviets, by surrounding their borders with an Iron Curtain, gave themselves the equivalent of American geographic isolation and thereby decreased the Soviet vulnerability to surprise attack and to atomic retaliation. Since strategic air power cannot fulfill its assigned mission unless it has knowledge of the disposition of the enemy's forces, secrecy was to become a major component of Russian strategy.

Simultaneously with an acceleration of its efforts to achieve equality with the United States in nuclear armament, the Soviet Union sought to compel the United States to disarm itself atomically. Henry Kissinger, in his book *Nuclear Weapons and Foreign Policy*, gives an excellent analysis of the Soviet propaganda attempt to outlaw nuclear weapons.[31] By

[31] Harpers, 1957, Chapter Eleven. For a history of Soviet nuclear disarmament policy see Joseph Nogee, *Soviet Policy Toward International Control of Atomic Energy*, University of Notre Dame Press, 1961.

many varying means the Soviet Union commenced its effort to make it impossible for the United States to utilize nuclear weapons.

Conclusion

The scientists who left Los Alamos, Oak Ridge, and the Metallurgical Laboratory to advocate the international control of atomic energy drew the attention of the United States to the implications for its security of this new power. Although many spoke optimistically of the elimination of war from the face of the earth and of the creation of world government, their immediate interest as reflected in their support of the Baruch Plan was to prevent a nuclear arms race. They believed that every reasonable man would be led by the "logic of facts" to the support of the plan.

The "facts" from which the scientists had drawn their "logical" conclusion that the Baruch Plan was *the* solution to the problem of nuclear weapons were viewed by them as being solely technical in nature. For this reason they regarded the Baruch Plan as a technical solution which any reasonable nation, given the facts, could be expected to support. Underlying the Baruch Plan, however, were political assumptions shared by American scientists and American political leaders which the Soviets could not accept. The emphasis on openness, the possibility of a Western veto over Soviet atomic industry, and the continued United States hegemony over nuclear matters among other things were contrary to the Russian view of their national interests.

Although the Baruch Plan would have committed the United States to a limited world government, it was consistent with the traditional political and military objectives of the United States. Nevertheless it actually did contradict the evolving relationship of the political commitments to the military capabilities of this nation. An implementation of the Baruch Plan would have interfered with the new American commitment to defend Western Europe against Soviet aggrandizement. Unless the United States was prepared to commit a vast army to defend Western Europe, it had no choice but to deal with the Russians from a position of nuclear strength. It was this American nuclear threat, according to Winston Churchill, that had kept the Russians from advancing to the English Channel.

Thus American nuclear policy, even at this early period, contained the implicit dilemma mentioned in Chapter One. At the same time that American political leadership endorsed the total elimination of nuclear weapons as advocated by the scientists, it pursued a military policy of overdependence upon nuclear weapons.

As 1946 gave way to 1947 and 1947 to 1948 it became obvious to more and more scientists that they could not divorce the problem of the control of atomic energy from its political context. As tension rose in the world the problem of the atomic bomb became more obviously and intimately joined to the political issues that divided East and West. Whereas in 1946 Soviet "intransigence" had been viewed as resulting from its irrational suspicion of the capitalistic West, by 1948 scientists had begun to realize that the source of conflict was deeper than this. The Czech *coup,* the Berlin Blockade, and the onset of the Cold War made it evident to most scientists that the conflict between East and West was over the political future of Europe and Asia. For the United States and the scientists the question of what role if any the atomic bomb should play in the emerging political struggle became a matter of highest importance.

CHAPTER THREE

THE DEVELOPING SCIENTIFIC CLEAVAGE OVER THE COLD WAR AND THE HYDROGEN BOMB

The Cold War Begins: Fission of the Scientists' Movement (1947–1948)

THE growing awareness among scientists of the inconsistency in American nuclear policy between the goal of international control of atomic energy and that of containment of the Soviet Union by nuclear weapons resulted in the political fragmentation of the scientists' movement by 1950. The temporary unity of this movement in its attempt to achieve the international control of atomic energy disappeared, and in its place three major groupings of scientists with differing approaches to the problem of nuclear weapons emerged. The first division among the scientists appeared slowly during the years 1947–1948 as the United States began to rearm in the face of Soviet hostility. The second was to take place in 1949 over the issue of whether or not the United States ought to attempt the "crash" development of the hydrogen bomb.

Although scientists had been in accord with official American policy toward atomic weapons as it was formulated in the Baruch Plan, over-all United States foreign policy soon began to seem schizophrenic to scientists. When the Cold War began with a competition between the United States and the Soviet Union to determine the future of Europe, the United States faced the serious dilemma of wanting simultaneously to relieve itself of the potential threat of surprise attack by atomic weapons (as the Baruch Plan was designed to do) and also to keep devastated Europe free from domination by the Soviet Union. Therefore the United States tried to achieve the former goal by continued support of the Baruch Plan for international control while at the same time obviously utilizing its atomic arsenal as a deterrent to Soviet aggression.

Furthermore, although the military supported the Baruch Plan, it was preparing tactics and strategy for a world where atomic energy would not be controlled.[1] And, while the Baruch Plan remained official American policy, a growing group within the military and both political parties was reaching the conclusion that the United States should base its security almost wholly on a policy of strategic atomic bombing.

When it could no longer be denied that the Baruch Plan was, at least for the time being, a failure, a feeling began to grow among a significant number of scientists that a new approach to the problem of atomic energy was required. Such a new approach would have to take into account those political roadblocks that had thus far prevented attainment of any agreement on the international control of atomic energy. At this same time, however, the writings, correspondence, and discussions of the scientists reveal that they were increasingly indecisive concerning the way in which scientists could accomplish this goal. After a survey of scientists, Richard Meier reported in 1948 that a sense of frustration had invaded the movement for international atomic energy control.[2]

Although a small minority of scientists continued to emphasize international control, the great majority of politically active scientists began to believe that the military containment of the Soviet Union had to take precedence, at least for the time being, over the search for the international control of atomic energy. On the basis of this difference in views the latter grouping will be referred to in this study as the "containment school" while the former will be called the "control school." While there are no hard and fast lines dividing these schools of thought from one another, the scientists who have actively participated in the debate over the issues of

---

[1] See Bernard Brodie, "War Department Thinking on the Atomic Bomb," *Bull. Atom. Sci.,* Vol. 3, No. 6, June 1947, pp. 150–55ff.

[2] Richard Meier, "What Should the Atomic Scientists Do Now?" *Bull. Atom. Sci.,* Vol. 4, No. 3, March 1948, pp. 81–82.

atomic weapons have tended to fall into one group or another.

Even as the manifestations of the Cold War multiplied with the establishment of Soviet control in Eastern Europe, a minority of scientists continued to believe that a short-range American policy determined by military expediency should not be permitted to undermine the possibility of achieving the vitally needed international control of the atom. Such scientists as Philip Morrison, Linus Pauling, Harlow Shapley, and a number of others then in the leadership of the Federation of American Scientists believed that the danger of uncontrolled atomic energy was the essence of the political division between the United States and the Soviet Union rather than a result of it. They felt that the conflict between the United States and the Soviet Union was actually generated by the existence of atomic weapons since both nations were insecure in a world with uncontrolled atomic energy; consequently both East and West were viewed as seeking to improve their own positions by creating military alliances. According to this view the division in the world would only be deepened by rearmament and the building of a Western alliance. It was argued that national energies should be devoted instead to the creation of an atmosphere of mutual confidence in which atomic energy could be controlled.

Accordingly, this group believed that the primary task of scientists should be to continue to devise new methods of control and to press them upon the nations. These scientists therefore suggested numerous schemes to end the arms race such as an international convention of physicists pledging abstention from military research. And, in general, they all attacked the "inflexible" American position and criticized the "intransigence" of the United States in continuing to present the obviously unsuccessful Baruch Plan; they wanted instead a first step toward disarmament which would open the way for more elaborate measures to control the arms race.[3]

The scientists who thought this way concluded that it was their task to accelerate the education of the leaders and the peoples of the world to the facts of the atomic age. Man must be taught that all men share one interest above all others, the mutual interest in survival and security in a world where the atomic bomb could bring sudden death. "If scientists have any right to speak out in these times, they have because they know that the world of men and of nature is indivisible. It ought to be the task of science to lead the badly divided world to a unity of toleration, to understanding through cooperation, and not to close doors, to hole in and to start work on the bigger bombs."[4]

In contrast to fellow scientists whose views had been modified at least temporarily by the Cold War, these scientists continued to believe that one could continue to approach the problem of atomic weapons as one solves a problem in science and that, as all men desire the solution to the problem, the task of the scientists is only to devise a solution amenable to all. Hans Morgenthau has referred to this belief that the problems of politics can be solved by man's reason unaided by physical force as rationalism.[5] While not absent from the thinking of other scientists—or most twentieth century men for that matter—rationalism is most pronounced in the thinking of the control school scientists.

Essentially rationalism assumes that both evil and political conflict originate in man's ignorance. Men conflict with one another because they are ignorant of the greater bonds that unite them and do not realize that they can maximize the interests of all through cooperation. Consequently, the solu-

[3] The use of the term "first step" to designate this approach did not appear until after 1955. See Pauling's petition discussed in Chapter Five.
[4] Philip Morrison and Robert Wilson, "Half a World . . . and None: Partial World Government Criticized," *Bull. Atom. Sci.*, Vol. 3, No. 7, July 1947, p. 181.
[5] Hans Morgenthau, *Scientific Man Versus Power Politics*, The University of Chicago Press, 1946.

tion to the evil in the world and to the problems posed by politics is believed to be the growth and diffusion of knowledge.

Rationalism is a view whose origin lies within science itself, and it has gained credibility as scientific knowledge has eradicated many centuries-old problems. The efficacy of man's reason in controlling the external world through science has influenced many men to believe that reason can also control the very power released by science and that man, as he acquires knowledge, will learn to solve his worst problems, including war itself. In place of the method of war man can and should substitute the method of science as the means to solve his problems.

An excellent statement of this rationalist faith by a scientist is to be found under the newspaper headline, "Human Mind Held Able to Ban War," and is as follows: "The modern techniques of reclaiming knowledge, using businesstype machines and card machines to filter out facts and organize them are powerful techniques. Since the whole magnificent picture of evolution is the product of human brains, I cannot believe that human brains will not be able to find solutions to ways of influencing people that are superior to the old, inefficient and destructive way of beating them over the head, whether with clubs or atom bombs. . . . "[6]

This theme that the problem of war is due basically to a lack of social knowledge rather than to the nature of man himself or of the state-system is found in Pauling's proposal that the world establish " . . . a great research organization, the World Peace Research Organization." Here the world's experts " . . . would attack world problems by imaginative and original methods. . . . The time has now come for man's intellect to win out over the brutality, the insanity of war."[7]

[6] Statement of Ralph Gerard, *The New York Times,* November 28, 1959, p. 21.

[7] Linus Pauling, *No More War!,* Dodd, Mead and Company, 1958, p. 12.

Prior to the Soviet H-bomb the arguments of the control school carried little weight with politically active scientists. The political break between East and West and the continual threat of open war did not provide an environment conducive to the acceptance of the control school argument, and until the Cold War seemed to become less intense after 1955 the ideas endorsed by this group remained relatively ineffective.

THE CONTAINMENT SCHOOL

The majority opinion which was supported by the leadership of the politically active scientists including Harold Urey, Robert Oppenheimer, Edward Teller, Arthur Compton, Isador Rabi, James Conant, and Hans Bethe, was based on a harsher view of threatening contemporary events. These men maintained that the Cold War had originated with the aggressive policies of the Soviet Union which was motivated for ideological and nationalistic reasons to dominate at first Western Europe and then the world. In their view, Soviet expansion had been and was being made possible by American weakness and indecision; further deterioration of the Western military position could result in an imbalance which would tempt the Soviet Union to wage aggressive war. In the face of this situation the West, in the view of these scientists, had to contain Soviet aggression and the only effective deterrent to Soviet expansionism was the threatened use of American atomic weapons. Although it was possible that such American containment of the Soviet Union might indeed lead eventually to an atomic arms race, a policy like this appeared to most scientists to be a necessary expedient until the power vacuum surrounding the Soviet Union could be filled in other ways.

This group of men, which will be called the containment school of scientists, believed that the question of the political future of Europe had provided the incentive for the Cold War. They felt, too, that the policy of the United States should be to draw closer to Europe in all fields, including the military, helping to rebuild European military power, and participating in the reconstruction of Europe.

The General Advisory Committee (GAC) of the Atomic Energy Commission under the chairmanship of Robert Oppenheimer was the principal stronghold of this position. These men were responsible quite directly for the continuing strength of the United States in the field of atomic weapons and they felt that the security of the United States required continued military exploitation of atomic energy until some system of dependable controls was established. Later, in 1954, Oppenheimer was to summarize the feeling of the General Advisory Committee at this time (1946): "Without debate—I suppose with some melancholy—we concluded that the principal job of the Commission was to provide atomic weapons and good atomic weapons and many atomic weapons."[8]

Although the containment scientists stood united on the pressing need to rebuild a Western military force, differences existed among them on the matter of international control of atomic energy. The various positions taken by these men on this subject foreshadowed the lines along which they would eventually split at the time of the hydrogen bomb controversy.

One group of containment scientists agreed with Eugene Rabinowitch that, despite the intense antagonisms of the Cold War, the scientists' struggle to achieve international control should not be abandoned. They believed that the search for a control system should parallel military rearmament. Like the control school scientists, this group had a rationalistic faith that East and West, despite their political differences, could still find some technical method to control atomic weapons if they searched hard and long enough. In 1948 Rabinowitch wrote: "The main reason why the UNAEC [United Nations Atomic Energy Commission] is now terminating its deliberations is not the absolute and ultimate incompatibility of the two opposing points of view on atomic energy control. The

[8] U.S. Atomic Energy Commission, *In the Matter of J. Robert Oppenheimer,* U.S.G.P.O., 1954, p. 69.

crux of the problem facing the commission is the elaboration of effective *mechanisms,* not agreement on generalities; and it is by no means obvious that the present controversies would be insoluble if discussion were permitted to proceed on a *matter of fact* basis."[9]

At the other extreme in the containment school were those scientists like Edward Teller and Robert Oppenheimer who believed that the failure of the Baruch Plan represented, for the time being at least, the destruction of any realistic hope for international control. Both Teller and Oppenheimer believed that in the face of Soviet hostility the primary responsibility of scientists was to assist in the rebuilding of America's military capabilities. Furthermore, they feared that continued emphasis on nuclear disarmament might weaken the Western will to rebuild military strength. Speaking of his colleagues and their attitudes in mid-1946, Oppenheimer told the Gray Board: "I tried to explain to them that the jig was up [i.e., there was no hope for the Baruch Plan] because that was relevant to getting back to work. At the same time I could not come out and say 'This [the Baruch Plan] is a hopeless thing,' because I had some official connection with the Government until the Government had itself said so."[10]

Oppenheimer even went to the extent of advising the Administration that the United States ought to abandon the United Nations negotiations on atomic energy control. According to General Frederick Osborn, Deputy U.S. Representative to the United Nations negotiations on atomic energy control (1947–1948), Oppenheimer "felt certain that, if the Iron Curtain was not lifted, any plan of international control would be exceedingly dangerous to the United States. . . . [Oppenheimer feared] that if we continued these negotiations we

[9] Eugene Rabinowitch, "The Narrow Way Out," *Bull. Atom. Sci.,* Vol. 4, No. 6, June 1948, p. 185. (Italics mine.) As we shall see, this notion asserts itself strongly after 1955 in the argument of scientists for *technical* talks on a first step agreement for a nuclear test ban.

[10] U.S. Atomic Energy Commission, *In the Matter of J. Robert Oppenheimer, op.cit.,* p. 45.

would make some compromises which without our fully realizing it would put us in the position of having accepted an agreement for the control of atomic energy, possibly with prohibition of bombs, without in reality the Russians having lifted the Iron Curtain. . . . [This] would put the United States in a very dangerous position of not really knowing what was going on in Russia, whereas the Russians would know all about what was going on here."[11]

These scientists, especially Oppenheimer, were particularly skeptical concerning the sources of Soviet conduct. They doubted very much that the Soviet Union's actions arose from an insecurity created by the American monopoly of atomic weapons. Instead, they believed the causes of the conflict between East and West originated in the very nature of the Soviet system. In a speech to the National War College on September 17, 1947 Oppenheimer summarized his understanding of the origin of the East-West conflict: "The second aspect of our policy . . . is that, while these proposals [the Baruch Plan] were being developed and their soundness explored and understood, the very bases for international cooperation between the United States and the Soviet Union were being eradicated by a revelation of their deep conflicts of interest, the deep and apparently mutual repugnance of their ways of life, and the apparent conviction on the part of the Soviet Union of the inevitability of conflict—and not in ideas alone, but in force."[12]

It can be seen that the containment school of thought possessed within itself a latent conflict over the question of the need to pursue the international control of atomic energy simultaneously with rearmament. This conflict was seriously to divide these scientists when they considered the issue of the hydrogen bomb. On one side of the issue would be ranged those scientists who favored another try to achieve control; on

[11] *ibid.*, p. 344.
[12] *ibid.*, p. 43.

the other side would be those who believed any such attempt would be unwise. However, when this moment of decision arrived, Oppenheimer and his associates on the GAC would shift to the position that the search for control of atomic energy must accompany nuclear rearmament.

The H-Bomb Issue: Cleavage Within the Containment School

At the same time that the scientists' movement was dividing into two schools of thought on the issue of control or containment, a second development was taking place. Whereas the debate over the Baruch Plan had been largely public in nature, the continuing debates on national policy toward atomic weapons were being isolated behind closed doors due to the requirements of military security.

While political agitation still dominated the scientists' journals such as the *Bulletin of the Atomic Scientists,* the scientists were playing their most vital political role more and more in secret.[13] In consequence, significant participation in the debate was restricted to those scientists actually behind the closed doors and in particular to the men in the containment group who held advisory, administrative, or research positions in the government.

BACKGROUND FOR THE H-BOMB CONFLICT

The position of the General Advisory Committee. The most important members of the containment school, the scientists on the General Advisory Committee (GAC) of the Atomic Energy Commission, shared with their colleagues in the control school a largely repressed concern over the irrationalism

---

[13] The only view into this world is provided by the report of the Personnel Security Board (Gray Board) of the Atomic Energy Commission on the security clearance of Robert Oppenheimer entitled *In the Matter of J. Robert Oppenheimer.*

of American national security policy.[14] They were well aware of the vacillation between the pursuit of international control of atomic energy and the dependence solely upon the combination of strategic air power and atomic bombs for deterrence of Soviet expansion.

The GAC scientists were equally disturbed over the immediate problem of the American military posture; they believed that American total dependence upon strategic bombing was extremely unwise. Not only was it believed by them that this policy would be insufficient to meet the character of the Communist threat but also that it would become exceedingly dangerous when the Soviet Union also possessed the atomic bomb. Unless America prepared for this eventuality, it would be reduced to just two alternatives in the face of Soviet piecemeal encroachment against the Free World—total nuclear war or capitulation.

These scientists in the containment school believed that they had a responsibility to discover another alternative which would enable men everywhere not only to survive but to preserve their freedom also. An excellent expression of this view is contained in excerpts from a letter written in 1948 by General Advisory Committee Chairman Robert Oppenheimer to the chairman of the General Board of the Navy. Although Oppenheimer concedes that it is an expression of his own thoughts as chairman of the Advisory Committee, he states

[14] The General Advisory Committee of the Atomic Energy Commission consisted of the following in the fall of 1949: Dr. J. Robert Oppenheimer, Director, Institute for Advanced Study, Princeton, N. J., Chairman; Dr. James B. Conant, President, Harvard University, Cambridge, Mass.; Dr. Lee A. DuBridge, President, California Institute of Technology, Pasadena, Calif.; Dr. Enrico Fermi, Professor of Physics, Institute for Nuclear Studies, University of Chicago, Chicago, Ill.; Dr. I. I. Rabi, Chairman, Department of Physics, Columbia University, New York, N. Y.; Hartley Rowe, Vice President and Chief Engineer, United Fruit Co., Boston, Mass.; Dr. Glenn T. Seaborg, Professor of Chemistry, University of California, Berkeley, Calif. (Seaborg did not attend the October 29th meeting); Dr. Cyril S. Smith, Director, Institute for the Study of Metals, University of Chicago, Chicago, Ill.; Oliver E. Buckley, President, Bell Telephone Co., N. Y.

that "it may give some background for what we started out
to do and what we did do in the descriptions we gave in the
General Advisory Committee." For this reason the inclusion
here of the entire statement is worthwhile:

"Whatever our hopes for the future, we must surely be prepared, both in planning and in the development of weapons,
and insofar as possible in our 'force in being,' for more than
one kind of conflict. That is, we must be prepared to meet the
enemy in certain crucial, strategic areas in which conflict is
likely, and to defeat him in those areas. We must also be prepared, if need be, to engage in total war, to carry the war to
the enemy and attempt to destroy him. One reason why we
must keep both of these objectives in mind (and they call for
quite definite plans and quite different emphasis as to equipment, troops and weapons) is that it may not be in our hands
to decide. With this reservation, it seems appropriate to suggest
that there may be two phases to the problem.

"At the present time [1948], to the best of my knowledge,
the Soviet Union is not in a position to effectively attack the
United States itself. Opinions differ and evidence is scanty as
to how long such a state of affairs may last. One important
factor may be the time necessary for the Soviet Union to carry
out the program of atomic energy to obtain a significant
atomic armament. With all recognition of the need for caution
in such predictions, I tend to believe that for a long time to
come the Soviet Union will not have achieved this objective,
nor even the more minor, but also dangerous possibility of conducting radiological warfare.

" . . . Insofar as the United States need not for some time
to come fear a serious and direct attack on this country, it
would seem to me likely that our primary objective would be
to prevent the success of Soviet arms and Soviet policies, to
carry out a policy of attrition, and not to engage in a total war
aimed at destroying entirely the sources of Soviet power. There
are many arguments for this and I have little to add to the

obvious ones. Yet, the general political consideration that the consequences, even in victory, of a total war carried out against the Soviet Union would be inimical to the preservation of our way of life, is most persuasive to me.

"On the other hand, as the time approaches, if it ever should, where as a result of political or military success in Europe or Asia, as a result of advancing technological development and improved industrial output, the Soviet Union becomes a direct threat to the United States, we shall no longer have this option. We should no longer have this option if the maintenance of a strategic area such as Western Europe or Japan could not be achieved without a direct attack on the sources of Soviet power.

"From this it seems to me that two conclusions would seem to follow: (1) That we must be prepared, in planning, in logistics, and in development, for more than one kind of war; and (2) that the very greatest attention must be given to obtaining reliable information about the state of affairs within the Soviet Union bearing on its military potential.

"One final comment: There is to my mind little doubt that were we today, with the kind of provocation which the Soviet Union almost daily affords, to attack the centers of Soviet population and industry with atomic weapons, we should be forfeiting the sympathy of many potential allies on whose cooperation the success of our arms and the fundamental creation of a stable peace may very well depend. These same people would no doubt be almost equally disturbed were we to renounce, irrespective of the development of Soviet power, recourse to such armament."[15]

In this statement four fundamental beliefs shared by the GAC scientists stand out as the basis for their subsequent actions: (1) they rejected the doctrine of strategic air power except as a means to deter direct large-scale Communist attack

[15] *ibid.*, pp. 46–47.

on the United States, Western Europe, or Japan;[16] (2) they maintained that the United States must be prepared to wage wars ranging from limited to total; (3) they held that there would be a most critical need to possess a capacity for graduated deterrence when the Soviet Union achieved nuclear parity with the United States; (4) yet they believed that the Soviet Union would not achieve nuclear weapons "for a long time to come."

Whereas there had been hope earlier that the Baruch Plan might succeed in changing the nature of Soviet society, these scientists now believed that the United States could not change Soviet aggressive policy, and that therefore the nation must have a military capacity to deter implementation of Russian policy. They believed that then a Soviet inability to satisfy its ambitions would dampen the élan of Communist ideology and would give rise to internal counter forces which would in time cause the Soviet Union to evolve into a peaceful nation which would accept the international control of atomic energy. However, until this occurred it was imperative that the United States have adequate military power to meet all possible Soviet threats.

Reactions of scientists to American political developments. Meanwhile, internal political developments in the United States indicated to these scientists that the American public would be little inclined to fulfill their policy recommendations. Each plunge in the Cold War temperature convinced more and more Americans of the need to wrap strict security precautions around United States atomic "secrets" and to talk ever more belligerently about the "stockpile of bombs." Increasingly the atomic bomb was believed by many political leaders to be the only defense against invasion of Europe by the new Tartar hordes.

---

[16] As early as 1946, Oppenheimer wrote of atomic weapons, "they are not policy weapons but . . . are themselves a supreme expression of the concepts of total war."

Fortunately this emphasis on nuclear containment was complemented by the establishment of the North Atlantic Treaty Organization (NATO) in 1949. In theory at least, the West would thus have a shield against minor aggression and a trip wire to trigger nuclear retaliation. Nevertheless, the Western military posture was to remain one which was almost totally dependent upon nuclear weapons.

Just as the scientists had predicted, moreover, the Cold War and the arms race appeared to be harming American democracy. Each deterioration in the international situation led to greater emphasis on the atomic weapon as the answer to the Soviet threat and to demands for more stringent internal security methods by which to protect American atomic secrets. Such measures were strongly advocated by an important group with isolationist tendencies which had emerged early in the Cold War. This group which was well represented in both major political parties held that the two fundamental tasks for the United States were to arm itself with atomic weapons for strategic warfare and to protect the "secret of the bomb." While these neo-isolationists appreciated the external challenge of communism, they believed that the more serious threat to American society was internal. The Air Force and its atomic armament could be counted upon to keep external communism at bay; therefore it was the internal threat of subversion and socialism that constituted the real danger to the Republic.[17]

Simultaneously with this political development the complementary military doctrine of strategic air power was becoming stronger in the military establishment. The affinity of the views of the isolationist politicians and certain Air Force officers resulted in a tacit alliance between them. It became inevitable that the containment school of scientists which supported a balanced military establishment would clash with this alliance.

As early as the 1920's the impact of air power on warfare

[17] For an excellent discussion of this type of thinking see Norman Graebner, *The New Isolationism*, Ronald Press Company, 1956.

had convinced certain military thinkers that offensive air power was invincible. This view had been reinforced by the facts of the atomic age which had been elucidated by the scientists. The "logic of the facts" seen by these military thinkers confirmed their conviction of the decisive nature of strategic bombing. As "the men who built the bomb" had themselves declared, the atomic bomb assured the decisiveness of surprise and of the offensive; against strategic atomic bombing science could provide no defense.

Specifically, the Air Force doctrine of strategic air power promised to counter the three advantages which the Soviet Union seemed to have over the United States; (1) the closed, secretive nature of Soviet society as opposed to the open democratic nature of American society, (2) Soviet control of the world's heartland, and (3) huge reserves of military manpower. Given the territorial expanse and the huge population of the Soviet Union, the Air Force argued that the Communist military challenge could only be met by atomic weapons. Not containment of communism at its periphery but the threat to destroy Russia itself was the solution to the problem of Communist aggrandizement.

The appeal of such Air Force doctrine to the neo-isolationists was reinforced by the latter's general belief that the "basic Soviet strategy" was to "bleed the United States" economically. These persons feared that by causing the United States to maintain a vast, expensive military establishment and to pour its resources into wars of attrition the Soviet Union would cause the bankruptcy and socialization of the American economy. The Air Force doctrine of strategic bombing offered an economical way to meet the Soviet threat. Long before the Eisenhower Administration introduced the "New Look" which, as Walter Millis described it, promised to substitute "gadgets for bloodshed in war," persons like Truman's Secretary of Defense Louis Johnson were intrigued by the prospects of containing the Soviet Union at a minimal expense. In particular this view delighted the neo-isolationists who detested "big government" and involvement in world affairs only slightly less than they hated communism.

As the Cold War grew worse, this neo-isolationist group believed more and more strongly that the most fundamental American task was to maintain the "secret of the bomb." If necessary, some argued, the whole field of atomic physics should be returned to military control; in any event they maintained that scientists had to be watched as they constituted the main threat to the American "secret." Thus it was that in 1948 Edward Condon, who was then Director of the National Bureau of Standards, was singled out as one of the "weakest links" in the chain that bound the "secret" of the American atomic bomb. Soon, other candidates for this appellation were ferreted out by various legislative committees. While the revelation of a Soviet atomic espionage network in Canada provided confirmation that there really was sufficient cause for concern over espionage, a widespread suspicion of scientists in general was a very bitter pill indeed for the scientists to have to swallow.

In 1949 the antagonism between scientists and the proponents of American neo-isolationism reached a new peak in a clash over two specific issues in mid-year. The issues were (1) Senator Bourke Hickenlooper's attack on Lilienthal's "incredible mismanagement" of the AEC, and (2) the requirement of an FBI investigation of each recipient of an AEC fellowship.

The effect of this clash was to convince many scientists of a threat to the freedom of American science inherent in the Cold War. In 1945–46 scientists had succeeded in their efforts to achieve American commitment to civilian control of atomic energy domestically and to a quest for international control externally. Scientists had considered both of these measures to be essential if America were to be secure and science were to remain free. They had believed, however, that even if the attempt to achieve control should not succeed, American well-being would remain dependent upon a free science. "Security through achievement" rather than through "concealment" was the watchword of these scientists. They felt that American basic science would slowly falter under continued military control and security regulations. Thus, while an attempt to regulate American science in the interest of security and secrecy would retard the Soviet attainment of an atomic capability only slightly if at all, it would surely destroy all American science until there soon would be nothing to conceal.

Now it seemed to these scientists such as Oppenheimer and his colleagues on the GAC that America was mistakenly seeking to defend itself against communism through reliance upon secrecy, security investigations, and strategic bombing. Hanson Baldwin summed up this pursuit of total security which disturbed many scientists. Baldwin wrote as follows: "The attacks upon Lilienthal and the Atomic Energy Commission . . . stem fundamentally from the mistaken belief that secrecy is security, and from the naive and fallacious overdependence of Congress and the public upon the atomic bomb. The bomb has become, unfortunately, our psychological Maginot Line. . . ."[18]

The Soviet atomic bomb. Into this domestic atmosphere of suspicion burst the Soviet atomic bomb in mid-1949. Its effect was to reinforce both scientists' concern about where the politican would take the nation and politicians' suspicions of the loyalty of scientists. The Soviet achievement convinced many politicians that someone "had given the secret" to the Russians and this resulted in an intensification of the search for atomic spies. Meanwhile, some scientists erroneously believed that the nation, in its shock at this development, would now awaken to their Cassandra cries. As they had predicted, the Soviets had broken the United States monopoly on atomic weapons; the efficacy of the American nuclear deterrent was ebbing and Europe was still unable to defend itself against Soviet ground forces. The vital question which the nation could no longer avoid was what should and could be done to strengthen the military position of the West in the meantime.

[18] Hanson Baldwin, *The New York Times,* May 26, 1949, p. 5.

One possible answer to this question was offered by the fact that, even prior to the conception of the fission bomb, physicists had been aware of the great power released by the fusion of hydrogen into helium. In 1937 Hans Bethe had shown that fusion of hydrogen into helium through the so-called carbon cycle provided a large part of the energy output of many stars. During World War II the possibility of making a fusion, hydrogen, or thermonuclear bomb had been explored but rejected in favor of concentration on a fission weapon. For a variety of reasons the thermonuclear program was assigned a relatively low priority among the AEC's many early postwar projects.[19]

However the Soviet atomic explosion made it inevitable that such questions be asked as: Was the United States doing all that it could to perfect a hydrogen bomb? If not, should it do more? If so, what specifically ought the United States to do?

The answers to these questions were unequivocally clear to AEC Commissioner Lewis Strauss, who on October 5th sent a memorandum to his fellow commissioners proposing that the way to keep ahead of the Russians was to take a qualitative jump. In time, Strauss reasoned, the lead of the United States in fission weapons would be overcome by the Soviet Union. The only hope for retention of American nuclear superiority

[19] Teller and Fermi had suggested the possible initiation of a fusion reaction by an atomic bomb in April 1942. Morgan Thomas, *Atomic Energy and Congress,* University of Michigan Press, 1956, p. 87. For a discussion of the pre-1949 status of thermonuclear research see, U.S. Atomic Energy Commission, *In the Matter of J. Robert Oppenheimer, op.cit.,* pp. 711–12, 949–50. In summary, it was the general view of scientists that it was "a long-term undertaking requiring very considerable effort" and that the exploitation of fission weaponry held far more promise for the then immediate future.

If one searches the postwar writings of American scientists he will find few direct references to the possibility of a hydrogen bomb. This was due to what Rabinowitch has called a "conspiracy of silence." As a consequence one finds only vague references in scientists' statements such as Bohr's reference in 1944 to "new possibilities." The only public discussion of the H-bomb prior to the fall of 1949 was a book written in German by Hans Thirring, *Die Geschichte des Atombombe,* Vienna: Neves Osterreich Zeilungs-und-Verlags-Gesellschaft, 1946.

was the perfection of a new order of nuclear weapon: the thermonuclear or Super weapon as Edward Teller had called it. Strauss therefore proposed that the AEC request the advice of the GAC on *how* to proceed with the development of Super.

Independently of Strauss, two scientists at the University of California at Berkeley, physicist Luis Alvarez, and chemist Wendell Latimer, reached a similar conclusion on the need to proceed with Super. Alvarez, believing that the thermonuclear program had been neglected by the Atomic Energy Commission, felt that the United States must determine immediately whether or not the thermonuclear weapon was a theoretical possibility and that if it were possible the United States must construct it before the Soviet Union.[20] With this project in mind Alvarez approached Ernest Lawrence, director of the Berkeley Radiation Laboratory. Lawrence had already been reached by Latimer who also was concerned over the seemingly laggard pace of America's atomic weapons program.

Under the impetus of this dual stimulation, Lawrence contacted Edward Teller who was at Los Alamos working on the weapons program. The talk with Teller convinced Alvarez, Lawrence, and Latimer of their worst fears. Not only was the thermonuclear program inadequate but according to Latimer: "in the period between 1945 and 1949 we didn't get anywhere in our atomic energy program in any direction. We didn't expand our production of uranium much. We didn't really get going on any reactor program. We didn't expand to an appreciable extent our production of fissionable materials. We just seemed to be sitting by and doing nothing."[21] On the positive side, they learned from Teller that the thermonuclear project had a "good chance if there is plenty of tritium available."[22]

---

[20] U.S. Atomic Energy Commission, *In the Matter of J. Robert Oppenheimer*, p. 774.

[21] *ibid.*, pp. 658–59.

[22] *ibid.*, p. 775. Tritium is the hydrogen isotope of mass 3. A plant for its production was built in 1951–53 on the Savannah River in Georgia.

The three Berkeley scientists then went to Washington and other points in the East to convince the Administration and Congress that the Atomic Energy Commission program was amiss and to volunteer their services and those of the Radiation Laboratory for an effort to improve the United States military position through work on the hydrogen bomb. By the 24th of October, the plans of these men had reached the implementation stage, and the Berkeley trio had confidence from their trip East that they had convinced the necessary persons in Congress, in the military, and in the Atomic Energy Commission, including a majority of the members of the General Advisory Committee. The only two clouds on their horizon were that the AEC's reactor experts were skeptical of their plans and that, according to word from Teller, "Oppie [Robert Oppenheimer] was lukewarm to our project and Conant was definitely opposed."[23] Nevertheless, reassured by Lewis Strauss and some interested California politicians, they decided that their views should be presented before the forthcoming meeting of the General Advisory Committee at the end of October.

Meanwhile, Lewis Strauss was arguing the case for an accelerated thermonuclear program both within the AEC and the Administration in general. Although he encountered stiff opposition in the person of the chairman of the AEC, David Lilienthal, as well as other officials, Strauss succeeded in having the issue presented to the GAC for its consideration at its next meeting. Upon the GAC meeting in October, then, were centered the hopes and fears of the opposed parties on the question of the advisability of building the hydrogen bomb.

In anticipation of the October 29th meeting, both Oppenheimer and Conant were giving considerable thought to the questions posed by the Russian atomic explosion; for that matter, so were the other members of the General Advisory Committee. They were all struck by the fundamental question: Where was the nuclear arms race taking the United States and the world? This question appeared to require an answer

[23] *ibid.*, p. 782.

before a decision could be made upon development of the hydrogen bomb.

Four years earlier, in 1945, Conant and Oppenheimer had been condemned by the liberal press for acts of "collusion" with the "militarists" when they had stood apart from the vast majority of their scientific brethren and supported the allegedly militaristic May-Johnson Bill.[24] Both of them had been close advisors to General Leslie Groves who had been a villain of sorts to many scientists as well as to others. They had believed fervently that the most important task for the country was the safeguarding of its weapons program until a foolproof control system could be worked out and implemented.

By 1949, both Conant and Oppenheimer had become very disturbed about the trend of events both domestically and internationally. Internally there appeared to be a paralysis of clear thinking concerning a wise response to the Communist challenge. More and more the American posture seemed to be just to sit still on its "secret of the bomb" and to erect a security curtain around all scientific knowledge. The United States was failing to utilize its great resources to create the weaponry which would be needed when the monopoly of the atomic bomb was broken. Furthermore, the attack on scientists' loyalty and the thoughtless security restrictions imposed were harming the science upon which America's long-run security was dependent. And externally, while America had faced up in part to the Communist threat through the Marshall Plan and the Truman Doctrine, it had failed to appreciate the significance of Soviet conventional military power. In Europe a growing neutralism threatened to topple pro-Western governments. The total American reliance on the atomic bomb was causing these peoples to choose between Communism and atomic annihilation.

For Conant and Oppenheimer the Soviet achievement of the atomic bomb at such an early date underscored the bankruptcy of American policy; they believed that the proposed

[24] Oppenheimer had later withdrawn his support.

hydrogen bomb project would just be more evidence of this bankruptcy. Conant's rejection of the hydrogen bomb project was based on the belief that "[it] was supposed to be an answer to the fact that the Russians had exploded an atomic bomb. Some of us felt then . . . that the real answer was to do a job and revamp our whole defense establishment, put in something like Universal Military Service, get Europe strong on the ground, so that Churchill's view about the atomic bomb would not be canceled out . . . this was sort of a Maginot Line psychology being pushed on us."[25]

Conant was unalterably opposed to a crash program to construct the hydrogen bomb as the answer to the Soviet atomic explosion. Therefore, in anticipation of the October 29th meeting of the General Advisory Committee, Conant contacted Oppenheimer whom he knew would at least be receptive to his viewpoint. He found Oppenheimer undecided; although Oppenheimer shared Conant's strong feeling of concern over the possible implications of a decision to construct the hydrogen bomb, he, like his fellow physicists Bethe and Rabi, could not decide on a course of action. He wrote Conant prior to the October 29th meeting: "What concerns me is really not the technical problem. I am not sure the miserable thing will work, nor that it can be gotten to a target except by ox cart. It seems likely to me even further to worsen the unbalance of our present war plans. What does worry me is that this thing appears to have caught the imagination, both of the congressional and of military people, as *the answer* to the problem posed by the Russian advance. It would be folly to oppose the exploration of this weapon. We have always known it had to be done; and it does have to be done, though it appears to be singularly proof against any form of experimental approach. But that we become committed to it as *the way* to save the country and the peace appears to me full of dangers."[26]

[25] *ibid.*, p. 387.
[26] *ibid.*, pp. 242–43. (Italics mine.)

THE OCTOBER 29TH MEETING OF THE
GENERAL ADVISORY COMMITTEE

The responsibility of the scientists. The scientists who gathered for the GAC meeting on October 29th realized that a great responsibility had been placed in their hands. The primarily technical question originally posed by Strauss— *how* to proceed with Super—had by the time of the meeting become enlarged to one pregnant with political, moral, and strategic implications. Although there is controversy over whether or not the question was formally put to the scientists, it is evident from testimony before the Gray Board that the GAC was expected to answer the question of what the American response to the Soviet atomic bomb *ought to be.*

The GAC scientists attempted to view the problem of a response to the Soviet atomic explosion in the light of both the short-range and the long-range security of the United States. Their concern for the former was aptly stated by Isador Rabi a few years later in his testimony before the Gray Board: "Following announcement of the Russian explosion of the A-bomb, I felt that somehow or other some answer must be made in some form to this to regain the lead which we had. There were two directions in which one could look: either the realization of the super or an intensification of the effort on fission weapons to make very large ones, small ones, and so on, to get a large variety and very great flexibility."[27]

At the same time, the scientists were conscious of the long-range threat which nuclear weapons would present to the world if nations became too dependent upon them to guarantee their security. "I think," Oppenheimer told the Gray Board, " . . . we [the General Advisory Committee] thought we were at a parting of the ways, a parting of the ways in which either the reliance upon atomic weapons would increase

[27] *ibid.,* p. 452.

further and further or in which it would be reduced. We hoped it would be reduced because without that there was no chance of not having them in combat."[28]

These concerns of the GAC scientists must have been reinforced by the two individuals who briefed them on the political-military situation. Both George Kennan, the representative of the Secretary of State, and General Omar Bradley, Chairman of the Joint Chiefs of Staff, shared the scientists' anxiety over the imbalance in American military posture, and both opposed the over-reliance of the United States on the doctrine of strategic air power and favored a policy of containment based on a capability to wage limited war.[29]

The report of the General Advisory Committee reflects the complexity of the situation in which the scientists found their country and themselves, wherein the scientists felt a responsibility to recommend measures which would meet the new Soviet challenge but yet would not foreclose the possibility of an eventual agreement with the Russians on control over atomic weapons. This task of the scientists was complicated by their realization that an intra-administration conflict over national defense policy was taking place. On one side of the debate within the Administration were those persons who advocated greater reliance upon strategic bombing with nuclear weapons. On the other side were those who believed that the nation was sacrificing in the name of total security a balanced military capability which could meet Communist aggression without resorting to total war. Just the week before the GAC was to meet, these groups had clashed before the House Armed Services Committee on the now famous Air

[28] *ibid.*, pp. 250–51.
[29] The fact that these two particular individuals went to the GAC to brief it indicates its great prestige at this time. Kennan is, of course, famous for originating the containment doctrine which was in direct opposition to the strategic air power position. Bradley had expressed his views earlier in October in the *Saturday Evening Post;* in his article, "This Way Lies Peace," October 15, 1949, pp. 32ff, Bradley advocated that tactical atomic weapons should be sent to Europe as the primary means to contain limited Soviet advances.

Force B36 versus Navy aircraft carrier controversy (October 6–21, 1949).

The recommendations of the General Advisory Committee. The report of the GAC was divided into four parts: a letter of transmittal; the positive recommendations; an evaluation of the hydrogen bomb; and the two annexes containing the scientists' political views. Through this division of the report the scientists tried to establish a clear separation between their technical judgment and their political assumptions. Nevertheless, the report must be judged as a whole and in these matters of high policy one cannot easily separate, if it is possible to do so at all, the technical and the political components of advice.

The position of the GAC on the hydrogen bomb was that: " . . . [an] imaginative and concerted attack on the problem has a better than even chance of producing the weapon . . . [yet] it would be wrong at the present moment to commit ourselves to an *all-out* effort towards its development."[30]

Specifically, the GAC recommended against a crash program for the production of the device which had been proposed by Teller and supported by the Berkeley scientists. For, as Rabi pointed out to the Gray Board, the GAC view that there was a "better than even chance" for success was tantamount to saying that the proposal was too vague to warrant the concerted effort desired by its advocates.[31]

While making it clear in their technical evaluation that the proposed crash program might succeed, the GAC scientists were emphatic in their opposition to such a program to develop the hydrogen bomb. "We all hope," the GAC report read, "that by one means or another, the development of these weapons can be avoided. We are all reluctant to see the United States take the initiative in precipitating this develop-

[30] Report of the General Advisory Committee as quoted in U.S. Atomic Energy Commission, *In the Matter of J. Robert Oppenheimer, op.cit.,* p. 79. (Italics mine.)
[31] *ibid.,* pp. 454–55.

ment. We are all agreed that it would be wrong at the present moment to commit ourselves to an all-out effort towards its development."[32]

This negative position of the GAC scientists was based on a mixture of technical, political, strategic, and ethical considerations. These elements cannot be separated from one another except for analytical purposes because, as Oppenheimer told the Gray Board, such recommendations as this one on technical feasibility are "total views."[33] This must be kept in mind as this discussion proceeds; it is entirely unjust, however, to criticize the GAC scientists because non-technical assumptions were part of their total view, a consideration which, unfortunately, was appreciated neither by the Gray Board nor by certain scientists at the time of the Oppenheimer security hearings in 1954.

A crucial feature of the GAC recommendation was the fact that it was evaluating "a single design which was in essence frozen" technically. This thermonuclear device known as "the Super" had been conceived by Edward Teller and had been under study by him for a lengthy period. "The essential point" with regard to this device, Oppenheimer told the Gray Board, was "that as we then saw it, it was a weapon that you could not be sure of until you tried it out, and it is a problem of calculation and study, and then you went out in the proper place in the Pacific and found out whether it went bang and found out to what extent your ideas had been right and to what extent they had been wrong."[34]

Thus, in order even to test Teller's concept, an extensive development and production program would be required. In addition to questioning the wisdom of such a crash program by which the feasibility of Super could be evaluated, a number of the GAC members such as Enrico Fermi questioned the theoretical underpinnings of this particular device.

[32] ibid., p. 79.
[33] ibid., p. 80.
[34] ibid., p. 79.

The GAC scientists concluded, therefore, that there should be more calculations and theoretical studies on thermonuclear processes prior to the commencement of the experimental and production stages.

Yet even if Super should prove theoretically and technically possible, these scientists doubted whether the thermonuclear weapon would be worth the effort in terms of "blast effect per dollar" in comparison with fission weapons. Furthermore, if the thermonuclear program were given a higher priority, it would divert resources to a program of questionable value from many fission programs that were soon to reach the "payoff" stage. In particular, the members of the GAC had in mind a fission weapon then "in the works" which would have had a yield equivalent to 500,000 tons of TNT. Such a weapon, they believed, would meet any conceivable military requirements of the United States.[35]

These scientists also opposed the assignment of a higher priority to the thermonuclear weapons program at Los Alamos than it already had, although they did not rule out the possibility of supporting a stepped-up program to develop a thermonuclear weapon at a future date. Instead of a crash program on the hydrogen bomb, these scientists recommended in the main section of their report an increased fission program of "weapons expansion, weapons improvement and weapons diversification."[36] This was essentially the same expansion of the nuclear weapons program that they had repeatedly advocated during the preceding few years. Like the thermonuclear program, however, diversification would require a great increase in the American atomic plant before the required neutrons could be made available. Furthermore, "new types of plant . . . would give a freedom of choice with regard to weapons."[37]

The tight budgets of the pre-Korean War period had made

[35] *ibid.*, p. 400.
[36] *ibid.*, p. 77.
[37] *ibid.*, pp. 77–78.

such an expansion of plant to refine or make fissionable materials impossible. Furthermore, as Oppenheimer himself admitted later, Latimer had been correct in his assertion that the General Advisory Committee had thus far been conservative in recommending expansion of the reactor program. It had been torn two ways in its policy recommendations in this area. At the same time that it advocated a general expansion of plant to increase the supply of fissionable materials, it had been reluctant to recommend specific costly reactors or piles which, due to technical advance, would soon become obsolescent. However, now that the General Advisory Committee had been given the go-ahead to propose a program, a risk did have to be taken.

The GAC now recommended that new reactors be constructed. Such plants would not only provide the neutrons for an increase in the fission program but, it was believed, if the Administration should decide at some time in the future to proceed with the thermonuclear bomb, would be able to supply the necessary materials for this also. These new facilities would later be supplemented by a heavy water reactor which was to be constructed on the Savannah River in Georgia.[38]

In the letter of transmittal and in the two annexes, the scientists on the GAC sought to state explicitly their political, military, and ethical views. Despite differences in emphasis as reflected in the two annexes the scientists essentially made three points: (1) they pointed out the dangers as they saw them inherent in the imbalance of American military power based on total reliance upon the strategic employment of nuclear weapons; (2) they argued that nuclear weapons posed a serious long-term threat to American security; and (3) they proposed that the President inform the American people and the world of the nature of these two dangers to American

[38] *ibid.*, p. 457.

and world security. These three points will appear again and again in the discussions of scientists over nuclear weapons in the succeeding years.

The concern of these scientists over the imbalance in American military power was reflected in their recommendation that the fission program be expanded and diversified so as to provide the United States with tactical nuclear weapons to employ in its own defense and in that of its allies. These men did not believe, however, that tactical nuclear weapons alone would constitute a sufficient alternative to strategic bombing. Conventional military power had to be increased as well. "I was very aware of the fact," Oppenheimer told the Gray Board (and it was an opinion his GAC colleagues shared), "that you couldn't, within the atomic energy field alone, find a complete or even a very adequate answer to the Russian breaking of our [atomic] monopoly."[39]

At the same time that the scientists advocated increased reliance upon tactical nuclear weapons as well as on conventional forces, they nevertheless sought to convey to the Administration their conviction that the nuclear arms race would be ultimately detrimental to the United States. In particular, they tried to point out the danger to the United States which the development of thermonuclear weapons would present. The advent of this type of weapon would increase American vulnerability to surprise attack, accelerate the trend away from a balanced military program, and give a false sense of security.

The minority and majority reports did differ, however, in emphasis on the ethical and political components of their respective appeals against the hydrogen bomb. The minority annex, signed by Fermi and Rabi, stressed the ethical problem involved; the majority annex, signed by Oppenheimer, Conant, DuBridge, and Buckley, was, on the other hand, more political in its emphasis:

[39] *ibid.*, p. 86.

Minority Annex: "The fact that no limits exist to the destructiveness of this weapon makes it[s] very existence and the knowledge of its construction a danger to humanity as a whole. It is necessarily an evil thing considered in any light. For these reasons, we believe it important for the President of the United States to tell the American public and the world that we think [it] wrong on fundamental ethical principles to initiate the development of such a weapon."[40]

Majority Annex: "In determining not to proceed to develop the super bomb, we see a unique opportunity of providing by example some limitations on the totality of war and thus of eliminating the fear and arousing the hope of mankind."[41]

The scientists who made these statements felt that the President should educate the American people and the world to the dangers for mankind inherent in thermonuclear weapons and the nuclear arms race. They were convinced that if the United States took the symbolic step of refusing to construct such weapons and declared them to be immoral, mankind would be spared a threat to its very existence. Whether by direct negotiations with the Soviet Union as proposed by the minority annex or by foreswearing American initiation of the weapon as the majority annex favored, the scientists on the GAC hoped to prevent an accelerated thermonuclear arms race.

Yet, if such an attempt to prevent the development of the hydrogen bomb by the Soviet Union should fail, then, these scientists believed, the United States would have no choice but to initiate a high priority thermonuclear program. In the meantime research should, of course, continue on thermonuclear problems; however, the primary emphasis of the AEC should be on the expansion and diversification of fission weapons until atomic weapons were eliminated through international agreement.

[40] *ibid.,* pp. 79–80.
[41] *ibid.,* p. 80.

THE OUTCOME OF THE GENERAL ADVISORY
COMMITTEE REPORT

The recommendation of the GAC against a crash program on the hydrogen bomb which was passed on to the President by the AEC with the separate comments of the commissioners caused great consternation among the military, scientific, Congressional, and Administration proponents of the H-bomb and gave rise to an intense political struggle. Senator Brien McMahon, Chairman of the Joint Committee on Atomic Energy, was especially disturbed and personally appealed to President Truman for an acceleration of the thermonuclear program. In response to this reaction and in order to resolve the intra-administration conflict, President Truman appointed a committee composed of Secretary of State Dean Acheson, Secretary of Defense Louis Johnson, and AEC Chairman David Lilienthal to recommend whether or not the hydrogen bomb ought to be developed. In the meantime, the political struggle continued among the proponents and opponents of the bomb.[42]

Whereas both Acheson and Johnson favored the crash program on the hydrogen bomb, Lilienthal was in strong opposition. His initial reservations on the project had been considerably strengthened by the arguments of the GAC that the hydrogen bomb would only further imbalance the American defensive posture and that a decision on the hydrogen bomb should be postponed until there could be a thorough review of America's strategic position. Although both the Secretaries of State and Defense were agreeable to such a review, they feared it might be used to sidetrack an acceleration of the thermonuclear program. They therefore insisted upon an increased research pro-

[42] For the history of this larger political struggle see Samuel Huntington, *The Common Defense,* Columbia University Press, 1961, pp. 298ff; also see J. Stefan Dupré and Sanford Lakoff, *Science and the Nation—Policy and Politics,* Prentice Hall, 1962, pp. 114–23.

gram to examine the feasibility of the thermonuclear concept and so recommended to President Truman.

On the basis of the committee report, President Truman announced his decision on January 31, 1950. This decision which has been appropriately labelled a "minimal" decision[43] sought to end the intra-governmental debate over the hydrogen bomb without really settling the issue; the President apparently desired to delay a final commitment as long as possible. Thus, while he "directed the Atomic Energy Commission to continue its work on all forms of atomic weapons, including the 'hydrogen' or super bomb," President Truman did not specifically commit the United States to the crash program desired by the proponents of the hydrogen bomb. He did, on the other hand, in the conclusion to his announcement order the review of American defense policy advocated by the GAC scientists in their October 29th report.

In retrospect it appears that both sides in the hydrogen bomb debate enjoyed some success and only its opponents suffered any losses. In the case of the hydrogen bomb advocates, the President's decision provided the basis for a recommendation from the Secretary of Defense on February 24th that the United States prepare for the quantity production of hydrogen weapons without waiting for the experimental proof that a thermonuclear weapon was technically feasible. President Truman's approval of this proposal on March 10th gave the hydrogen bomb advocates the very thing they had wanted when the debate had begun in the preceding fall.[44]

At the same time the GAC scientists had at least succeeded in committing the United States to undertake a

---

[43] This is the apt characterization of the decision made by Warner Schilling, "The H-Bomb Decision: How to Decide Without Actually Choosing," *Political Science Quarterly*, Vol. 76, No. 1, March 1961. Professor Schilling has in preparation a detailed study of the hydrogen bomb controversy.

[44] *ibid.*, p. 44.

thorough review of national defense policy. And although both the GAC and Lilienthal had actually desired such a review to *precede* any decision on the hydrogen bomb, they had won at least the concession that it take place concurrently with the exploration of thermonuclear possibilities. Furthermore, as reflected in the President's announcement of January 31st, they had won the point that acceleration of the thermonuclear program should not be permitted to detract from progress in the fission program. Both these concessions to the GAC position were to be important in the months preceding and following the outbreak of the Korean War.

With respect to their third point that the President should inform the nation concerning the dangers of the nuclear arms race and that there should be national debate on this subject, the GAC scientists failed. The entire debate over the wisdom of the President's course of action was carried on in secret although there were frequent allusions in the press to the fact that the debates were in progress. With the exception of "leaks" to Drew Pearson and the Alsops, and a rather cryptic press statement by Senator McMahon on January 30, 1950, there had been only one serious breach of the security curtain. On November 1, 1949, Senator Edwin Johnson of Colorado appeared on a television program where he took advantage of his opportunity to chastise those scientists who were "forever advocating" that the United States "give away its secrets." Then the Senator volunteered, much to his subsequent chagrin: "Here's the thing that is top secret. Our scientists from the time the bombs were detonated at Hiroshima and Nagasaki have been trying to make what is known as a Super bomb. They want one that has a thousand times the effect of that terrible bomb—and that's the secret, the big secret that scientists in America are so anxious to divulge to the whole scientific world."[45]

---

[45] Quoted in Ralph Lapp, *Atoms and People*, Harpers, 1956, p. 106.

The Status of the Scientific Cleavage in 1950: Fission of the Containment School[46]

The debate over the hydrogen bomb brought to the surface differences within the containment school which still appear today. There is great misunderstanding concerning the nature of the cleavage among these scientists over the hydrogen bomb, and both personal vindictiveness and national strategy differences have been suggested as explanations of it. While these factors were indeed to become important in future issues on which members of the containment school would divide, they do not provide sufficient explanations of the conflict of the scientists over the hydrogen bomb.

THE FINITE CONTAINMENT SCHOOL

The position taken in the dispute over the hydrogen bomb by the General Advisory Committee will be identified in this study by the term "finite containment." According to this position it is both technically feasible and politically desirable to limit the nuclear arms race by international agreement at some *finite* point prior to the settlement of political differences between the United States and the Soviet Union while at the same time it is also necessary to *contain* Soviet aggression. In 1949 these scientists based this position on the belief that both of these countries had an overriding interest in preventing the development of hydrogen weapons and eventually in bringing all nuclear weapons under a system of international control.

Given the alternative of the risks of an infinite atomic arms race or those of a finite cut-off of that race, these scientists did and do believe that the greater probability of disaster lies with the former course of action. They hold, therefore, that at

---

[46] The control school is excluded from this summary because of its small role at this time. Its members made only a few feeble responses to the problem raised by the Soviet atomic explosion. They proposed a conference of the United States and the Soviet Union to discuss " . . . universal disarmament and an end to the Cold War." See Federation of American Scientists *Newsletter,* A753, November 30, 1949.

all times the United States should seek to arrest the arms race and believe that at some point both the United States and the Soviet Union will agree that it is to their mutual advantage to bring the arms race under control.

The members of the finite containment school believe that they have a responsibility as scientists to assist mankind to keep political developments abreast of technical advancement. In the hydrogen bomb controversy the General Advisory Committee desired to prevent a technical advance which would outstrip man's political ability to control it. They believed that a responsibility to present such advice was theirs because the atomic bomb had been their creation. As Oppenheimer, defending his actions at the time of the hydrogen bomb controversy, told the Gray Board in 1954: "I felt, perhaps quite strongly, that having played an active part in promoting a revolution in warfare, I needed to be as responsible as I could with regard to what came of this revolution."[47]

The scientists on the GAC sought to strike a balance between their desire to see an end to the nuclear arms race and their realization that American military power had to be bolstered. They reluctantly had come to the conclusion that tactical atomic weapons could bring about a refurbishment of American strength in limited war without closing the door to eventual international control of atomic weapons.

In the view of these scientists the decision to build the hydrogen bomb would have meant entering an entirely new world in which it might never be possible to bring nuclear weapons under control. With this decision it seemed probable that the nations, following the lead of the United States, would become totally reliant upon weapons of mass destruction and that nothing would then ever be able to persuade the nations of the need to eliminate them. Therefore, the finite containment scientists concluded that unless the development of the hydrogen bomb became utterly impossible to avoid, the United

---

[47] U.S. Atomic Energy Commission, *In the Matter of J. Robert Oppenheimer, op.cit.*, p. 959.

States ought to defend its interests only with the more manageable, yet still dangerous, fission weapons.

These scientists believed that the risk to which the United States exposed itself in trying to ban the hydrogen bomb prior to undertaking its development was a small one. Hans Bethe in the following excerpt from the Oppenheimer hearings expressed this view:

QUESTION: Would you describe briefly what you regarded as the alternative to going ahead with the thermonuclear program?

ANSWER: Yes. I thought that the alternative might be or should be to try once more for an agreement with the Russians, to try once more to shake them out of their indifference or hostility by something that was promising to be still bigger than anything that was previously known and to try once more to get an agreement that time that neither country would develop this weapon. This is enough of an undertaking to develop the thermonuclear weapon that if both countries had agreed not to do so, that it would be very unlikely that the world would have such a weapon.

QUESTION: Can you explain, Dr. Bethe, how you reconciled that view just described of wanting to make another try at agreement with Russia, with the view that you described a little while ago in which you expressed the feeling that negotiations with Russia on the A-bomb were hopeless?

ANSWER: Yes. I think maybe the suggestion to negotiate again was one of desperation. But for one thing, the difference was that it would be a negotiation about something that did not yet exist, and that one might find it easier to renounce making and using something that did not yet exist [than] to renounce something that was actually already in the world. For this reason, I thought that maybe there was again some hope. It also seemed to me that it was so evident that a war fought with hydrogen bombs would be destructive of both sides that maybe even the Russians might come to reason.

QUESTION: Didn't you feel that there was a risk involved in taking the time to negotiation [sic] which might have given the Russians the opportunity to get a head start on the H-bomb?

ANSWER: There had to be a time limit on the time that such negotiations would take, maybe a half year or maybe a year. I believe we could afford such a head start even if there were such a head start. I believed also that some ways could have been found that in the interim some research would go on in this country. I believed that also our armament in atomic bombs as contrasted to hydrogen bombs was strong enough and promised to be still stronger by this time, that is, by the time the hydrogen could possibly be completed, so that we would not be defenseless even if the Russians had the hydrogen bomb first.[48]

Support for Bethe's view also grew out of a strong conviction of the finite containment school that the Soviet Union was largely imitative in its atomic energy programs. "I believe," Oppenheimer stated in 1954, "that their [Soviet] atomic effort was quite imitative and that made it quite natural for us [on the GAC] to think that their thermonuclear work would be quite imitative and that we should not set the pace in this development. I am trying to explain what I thought and what I believe my friends thought."[49]

The finite containment scientists on the General Advisory Committee believed that an American decision not to construct the hydrogen bomb would again symbolize the sincerity of America's desire to end the atomic arms race. Many of these same scientists had also had this idea in mind when they had lobbied for a civilian Atomic Energy Commission and had supported the Baruch Plan. This same idea was to appear again after 1955 in the arguments for a nuclear test ban as a first step toward nuclear disarmament.

In the final analysis it would appear that the position of the

[48] *ibid.*, pp. 329–30.
[49] *ibid.*, p. 80.

finite containment scientists on the General Advisory Committee was dictated by their sense of responsibility and concern. Washington in late 1949 was a scene of confusion, drift, and acrimony. This was the period of the atomic spy scares, the fall of China, the B-36 controversy, and now the Russian atomic bomb. The character of the American reaction to the challenge of communism appeared to call for an appeal to reason. Whatever the merits underlying their own case, the scientists on the GAC sought to stem the tide of irrationalism which appeared to be engulfing the nation.

THE INFINITE CONTAINMENT SCHOOL

The infinite containment school of scientists agreed with their colleagues on the GAC that the principal rationale for the hydrogen bomb was to deter its use by others. However, they believed that this meant that construction should begin as soon as possible. Because they believed that the establishment of a control system over atomic energy was impossible in the foreseeable future, Edward Teller and the other scientists in this group argued that the nuclear arms race would necessarily continue and that therefore the United States should maintain its lead in that race by developing the H-bomb as quickly as possible.

Whereas the finite containment group took a position comparable to that taken by the Franck Report in 1945, the infinite containment school's position paralleled in one major respect that of Bohr's memorandum. In contrast to the finite containment school which believed that a feasible control system could be developed even in a politically divided world, the infinite containment school, like the Bohr Memorandum, argued that control over nuclear weapons would only be possible in a completely open world such as that envisioned in the Baruch Plan. Under the conditions of modern science the arms race would therefore be unavoidable until the political differences underlying that arms race were settled.

According to the infinite containment school a completely open world is the only alternative to security through armaments. Without an open world the pace of technical advance made possible by science makes any lesser system of control quickly obsolete.[50] Modern science and technology create novelties which cannot be anticipated, and there is no guarantee that a control system developed on the basis of an existing body of knowledge will be able to detect covert developments made possible by new knowledge. For this reason, these scientists argue, the attempt to control technical developments while secrecy still exists in the world is not only technically impossible but militarily dangerous.

This skepticism over the possibility of control stems from Bohr's thesis of the need for an open world if nations are to be secure; it had been expressed by Teller as early as 1946. Although he had been one of the most enthusiastic supporters of the Acheson-Lilienthal Proposals and the Baruch Plan, Teller had never believed that even they went far enough or fast enough, and a serious doubt had remained in his mind whether " . . . the control proposed in the Acheson Report . . . [was] sufficiently effective."[51] Teller had believed that the power of the proposed Atomic Development Authority ought to be increased in order to prevent any secret scientific developments. He had made the specific proposal that each nation should have the right to send as many agents as it pleased into any other; these agents should be treated as agents of the Authority with the right of access to *every* facility within the host nation. Furthermore, each citizen in the host nation should have both an obligation and the right under international law to inform these agents of any illicit atomic energy activities. Teller had indicated the heart of his position as follows: "One will not gain real confidence in the stability

[50] See Oppenheimer's earlier discussion of this with respect to the Baruch Plan in Chapter Two.
[51] Edward Teller, "A Suggested Amendment to the Acheson Report," *Bull. Atom. Sci.*, Vol. 1, No. 12, June 1, 1946, p. 5.

of the world structure until tyranny has disappeared from the earth and freedom of speech is ensured everywhere. To reach this goal may not be feasible in the immediate future. If the present proposal is put into effect, at least this much will have been achieved: We shall have a way to protect a man who has raised his voice for the purpose of safeguarding peace."[52]

Teller's belief in the need for openness as a prerequisite for a peaceful and secure world had been so strong that he suggested in 1946 that as soon as the nations had established the Atomic Development Authority proposed by the Baruch Plan the United States should declassify all its nuclear secrets. In exchange for these secrets the United States would create "an atmosphere of completely free discussion."[53] Such a situation, Teller would undoubtedly argue, was and is a necessary condition for world peace.

The Soviet Union's failure to accept the generous terms of the Baruch Plan and its subsequent erection of an Iron Curtain of secrecy convinced Teller and his colleagues in the infinite containment school that there was no alternative to the arms race and would be none until the Soviet Union should indicate by its actions a genuine willingness for peace. In effect, the nature of modern science, technology, and warfare made any hope to arrest the arms race a vain one until the Soviet Union became an open society.

This belief of the infinite containment school scientists is reinforced by their view that technology has largely replaced geography as the main element in national power. It is no longer he who controls the heartland who controls the world, but he who sets the pace in technological development. The lead-time that one nation has over another in the creation and application of knowledge may be decisive. In the energy-deficient world of the past a nation which was behind another in military technology could expend space to purchase time

---

[52] *ibid.*
[53] *ibid.*

in which to catch up. Traditionally, this has been the basis of American military policy.

Today, the infinite containment school has concluded, lead-time has become so important in a world of decisive weapons that any attempts to control continued development of weapons places a "competitive premium on infringement."[54] That is, in a world where nations through the applications of science are able to create effective novel weapons, a decisive military advantage flows to the nation which is first to possess such weapons. For this reason, under any system to prohibit technological advance in weapons, a temptation would always exist to violate the system and to achieve a technological coup which would shift the balance of power to one's advantage.

As a consequence of this "competitive premium on infringement" the only safe course for an open democracy is to forego any attempt to control technological development by international agreement. Whereas a democracy by its nature will not violate a control system, a totalitarian state could covertly infringe on the system. Even if such infringement should give the totalitarian state only a year's lead-time on a decisive weapon like the hydrogen bomb, this could be disastrous. On the other hand, it was reasoned, if both opposing camps proceed together in the development of these weapons mutual deterrence will maintain the peace. Thus these scientists believed that if the Soviet Union should achieve a monopoly of the hydrogen bomb the balance of power would shift in favor of the USSR. And regardless of what the United States chose to do in its weapons development, the vulnerability of the United States to hydrogen bomb attack provided the Soviet Union with more than adequate reason for it to initiate development of the weapon. Furthermore, disagreeing once again with the finite containment position, these scientists believed the United States would be incapable of countering Russian thermonuclear weapons with fission weapons. This

[54] John von Neumann, "Can We Survive Technology?" *Fortune*, Vol. 51, No. 6, June 1955, p. 152.

belief was expressed by John von Neumann in an extreme fashion in the following statement reported by Oppenheimer: "I believe there is no such thing as saturation. I don't think any weapon can be too large."[55] At the least, these scientists reasoned, a Soviet thermonuclear monopoly would give the Russians a psychological advantage over the West.

In summary, the scientists in the infinite containment school believed, as had the scientific advisory committee to the Interim Committee in 1945, that nuclear weapons could be made to serve the purpose of deterring war. Certainly they could not be eliminated unless the world became completely open; yet, these scientists felt, a positive good could come from the existence of nuclear weapons—the elimination of major war, and perhaps eventually of all war from the world.

Despite a stated position that the scientist's responsibility was simply "to explore and to explain,"[56] these scientists, too, operated on the basis of a sense of social responsibility. Their actions deviated greatly from Teller's view that "it is not the scientist's job to determine whether a hydrogen bomb should be constructed, whether it should be used, or how it should be used. This responsibility rests with the American people and with their chosen representatives."[57]

Indeed, they struggled as vehemently to convince the United States that it *ought* to develop the hydrogen bomb as the finite containment scientists did to convince American political leadership that it ought not to attempt this. The clash was between two opposed political views and not between the politically motivated and the scientifically motivated.

Teller's view that the scientists must do that which can be done does have its own inherent ethic. Implicit in his position is the belief that the advance of science is a great force for

[55] U.S. Atomic Energy Commission, *In the Matter of J. Robert Oppenheimer, op.cit.*, p. 246.

[56] Edward Teller, "The Work of Many People," *Science*, Vol. 121, No. 3139, February 25, 1955, p. 275.

[57] Edward Teller, "Back to the Laboratories," *Bull. Atom. Sci.*, Vol. 6, No. 3, March 1950, p. 71.

human progress and world peace. Despite the many dangers it presents, technological advance, including that in weaponry, provides mankind with the wherewithal to maintain peace by threat of retaliation and to eliminate the underlying social-economic causes of war.

The infinite containment scientists believe that if the peace-loving nations had a sufficient arsenal of atomic weapons they would destroy the will of aggressive nations to wage war. Specifically these scientists have argued that atomic weapons "would deprive the communist nations of their apparent major advantage, their huge manpower, by making it impossible for them to mass their troops for 'human sea' break-throughs. Possibly . . . [these weapons] would mean the end of all mass armies, conceivably even the end of major wars."[58]

In the view of these scientists nuclear weapons and the advance of technology pose a supreme paradox for mankind. The scientific revolution has brought man into the techno-political age and has enabled him to create weapons of such magnitude that a nation which possesses, even temporarily, a monopoly of new weapons of this type has a decisive military advantage. While the destructiveness of these weapons creates in mankind a strong disposition to eliminate them, their very decisiveness makes their elimination impossible save for a world under one government.

Conclusion

This brief history of the decision to develop the hydrogen bomb reveals both a failure on the part of American political leadership and the effect upon the scientist's advice of his sense of social responsibility. In effect, the Truman Administration turned to the scientists on the General Advisory Committee as its principal advisors on a major question of national policy, namely, what the American response should be to the Soviet atomic bomb. The scientists were asked whether

[58] Edward Teller as quoted in "Dr. Edward Teller's Magnificent Obsession," by Robert Coughton, *Life*, September 6, 1954, p. 66.

the nation should undertake an all-out program to produce a hydrogen bomb when the Administration itself had failed to develop a coherent, realistic policy toward nuclear weapons.

The scientists on the GAC accepted this responsibility and more. They tried to bridge the gap in political leadership and direction, and as a result it is not surprising that they were to be severely attacked for having exceeded their competence by those who sought to influence national policy in other directions. These critics have in general neglected to evaluate the actions of the GAC in terms of the fact that it had actually been asked for *policy* advice and that it gave it in a forthright, intellectually honest manner. For this reason such criticisms of this GAC action have been quite unjustified: the only question which ought to have been raised was whether or not the advice it gave was wise advice.

At a time when few others in government were concerned, the scientists on the GAC stressed the dangers inherent in the nuclear arms race. In retrospect it is difficult to disagree with their belief that the advent of thermonuclear weapons would be to the disadvantage of the United States and that the United States position would remain stronger if this development could be prevented. Yet, as Oppenheimer himself was to recount sadly, "what was not clear to us then and what is clearer to me now is that it probably lay wholly beyond our power to prevent the Russians somehow from getting ahead with it [the hydrogen bomb]."[59]

While it provided no sound alternative to the ultimate decision to build the hydrogen bomb, the GAC criticism of American military policy, on the other hand, did stimulate consideration of a very serious problem. Thus the GAC report provided the catalyst for a thorough review of American defense policy and no doubt thereby contributed to the improvement of the American military position at the outbreak of the Korean War.

---

[59] U.S. Atomic Energy Commission, *In the Matter of J. Robert Oppenheimer, op.cit.*, p. 80.

The GAC scientists did try to maintain a strict separation between their technical evaluation and their political views. In meeting the assignment of political leadership and in the exercise of their social responsibility they sought to point out to the Administration the broad nature of the problem presented by the Soviet bomb and to suggest an approach to its solution while simultaneously trying to provide an objective evaluation of the status of thermonuclear research. Nevertheless they were not completely successful in their attempt to separate their technical and political judgments, and the evaluation of the Super in the report of the GAC reflects the underlying political, military, and economic assumptions of these scientists. It was these assumptions of a political character and not simply technical judgments that divided the scientists over the hydrogen bomb.

Even though one must concede that the specific thermonuclear device under consideration was questionable, it is difficult to believe that this fact was controlling in the recommendation of the GAC against acceleration of the program. To paraphrase one scientist-proponent of Super, the members of the GAC could have at least recommended some additional support for the thermonuclear program. While their judgment on Teller's original device proved to be sound, this scientist argues, they knew that technical breakthroughs rarely come unless one is looking for them and that if the best minds of the country were brought in to concentrate on the problem, someone would find a solution to the theoretical problem if there were one to be found. And this is exactly what subsequently did happen.

The scientists favoring the hydrogen bomb felt particularly aggrieved because they believed that they had been betrayed by the members of the GAC whom they had regarded as their representatives to the government. The scientists on the GAC had had, in the view of the proponents of the hydrogen bomb, a responsibility to present the latter's views as well as their own in the report to the AEC. The failure of the GAC to do this was regarded by these scientists as an act of bad faith.

This sense of indignation on the part of the scientists favorable to the hydrogen bomb development was reinforced by their conviction that while they were true to the scientific tradition, their opponents on the GAC had been politically motivated. As one scientist who strongly supported the H-bomb development indignantly put it, the GAC scientists had based their advice on a "social theory" and not on the technical facts. Harold Urey referred to the position of the GAC on the bomb as based on a "curious prejudice."[60]

Little did the proponents of the hydrogen bomb realize that they too had a "social theory" in mind. Underlying their position were notions on the nature of the nuclear arms race and interpretations of possible Soviet action. For this reason the behavior of these scientists in attempts to achieve their goal also was brought into question by their scientific opponents. Writing of Lawrence and Teller, Oppenheimer regarded them as "two experienced promoters."[61]

Neither group of scientists appreciated the position in which the other found itself; instead they turned to interpretations which brought into question the integrity and propriety of the other's actions. The proponents of the bomb failed to realize that the GAC had been asked for its advice on the proper response to the Soviet atomic bomb and that the GAC was speaking as a political advisor rather than as a representative of the scientific community which should be judged by the objectivity of its advice. Nor was the behavior of the scientist-proponents of the hydrogen bomb out of order. Shut off from the regular channels of scientific advice, they had no choice, if their views were to be considered, but to appeal to Commissioner Strauss, Senator McMahon, and others. However, in this activity they were not solely intent upon seeking truth—no more so than were their opponents on the

[60] *The New York Times,* January 31, 1950, p. 10.
[61] U.S. Atomic Energy Commission, *In the Matter of J. Robert Oppenheimer, op.cit.,* p. 242.

GAC. Both groups were advocates of a strong political point of view.

Yet this conflict among the scientists served a very useful purpose; each group had a contribution to make to the formation of American policy. Without the scientist-proponents of the hydrogen bomb, President Truman and others might not have been stimulated to action subsequent to the negative advice of the GAC. And without the latter, the United States might not have undertaken its review of military policy which was to be so important when the United States remobilized after June 1950.

Nevertheless, these events had a deleterious effect on the morale of the scientific community and on the attitude of scientists toward the Administration. Scientists began to suspect one another's intentions and to question the capacity of the government to deal effectively with the problems of nuclear weapons. The scientists favorable to the hydrogen bomb project saw the action of their scientific representatives on the General Advisory Committee as an act which was irresponsible and unfathomable; the scientists who considered the ultimate problem to be that of the control of atomic weapons regarded their opponents' behavior in similar terms. And both sets of scientists developed a distrust of the government which is only now being dispelled. For the infinite containment scientists the lesson of the drama was contained in the slim margin of their success; they felt that only through a determined effort on their part had they been able to save the nation from the disastrous course proposed by their scientific opponents. Their opponents, on the other hand, began to wonder what hope there was for human survival if each new step in the atomic arms race were taken without prior attempts to negotiate a halt in that race. Despite the efforts of the GAC to convince political leadership of the imperatives of the atomic age the Administration had failed to review its nuclear policy in the light of the Soviet atomic explosion before making new decisions and had actually increased its dependence upon weapons of mass destruction.

CHAPTER FOUR

SCIENTISTS SEEK AN ALTERNATIVE
TO STRATEGIC BOMBING

ON JUNE 24, 1950, the fear of the General Advisory Committee members that an American strategic policy centered on weapons of massive destruction could not deter limited Communist aggression was realized. Less than a year following the Russian explosion of an atomic bomb and less than six months after the statement of Secretary of State Dean Acheson that South Korea was under the threshold of American retaliatory protection, the Russian satellite of North Korea invaded South Korea. After four years of containment the Communists had broken out. Now new questions had to be faced by the United States in its policy formation: Would the Communists strike elsewhere along their great periphery into the non-communist world? If so, where would this be? How could the United States stop them without risking atomic retaliation against its own friends or itself?

Fortunately for the nation, President Truman's coupling of an order to review American military policy with the decision to develop the hydrogen bomb had led to a thorough re-examination of that policy. In the resulting report completed in March 1950 and entitled *NSC68,* military officers, State Department officials, and scientists had recommended a vast increase in the United States' capabilities to wage limited war.[1] While this document did not rule out the use of tactical nuclear weapons, its main emphasis was on the need to increase America's conventional military power in

[1] Among the persons on the task force to draw up *NSC68* were Paul Nitze, George Kennan, and Robert Oppenheimer. For a discussion of *NSC68* see Samuel Huntington, *The Common Defense,* Columbia University Press, 1961, pp. 47ff. A detailed study of *NSC68* has been undertaken by Paul Hammond under the auspices of the Institute of War and Peace Studies of Columbia University and is to be published by Columbia University Press in 1962.

anticipation of the day when Russia could cancel America's nuclear deterrent.

In essence, *NSC68* recommended a major increase in overall American expenditure for armaments in order to meet various types of contingencies. The $13.5 billion military budgets of this period were cited as being far below the minimum required to meet American political commitments. The report recommended large increases in the military budget unparalleled in peacetime America. These recommendations might well have suffered an ignoble fate if it had not been for the outbreak of the Korean War. Instead, *NSC68* enabled the Administration to go before Congress at the beginning of the war with a ready-made plan for remobilization.

However, the crisis following the North Korean attack seemed to call for more drastic measures than just conventional remobilization. General concern over American capabilities heightened after Communist China entered the war in November 1950, and the fear developed that Russia would take advantage of the situation to strike elsewhere along the periphery of the Communist bloc. Among both military officers and such scientists as Robert Oppenheimer, Charles Lauritsen, and Jerrold Zacharias there quickly developed a belief that scientists should be mobilized to assist in the war effort.

Because of their positions within the government, the finite containment scientists tended to lead the attempts of scientists to influence United States security policy at this time. However, there were no significant differences of political opinion between the finite and infinite containment groups with the possible exception of opinion on the question of continental air defense. During this critical period the control school of scientists was of no real political significance.

Scientists Advise Development of a Capacity for Limited War: Project Vista

The advent of the Korean War, the initial American reversals, and the growing concern over the possibility of a general war had a profound effect on the scientists responsible for advising the government on nuclear weapons policy. As early as 1948, Robert Oppenheimer among others had argued for extensive investigation of the possibility of developing small atomic weapons and of utilizing them tactically. Both the General Advisory Committee to the AEC and the Atomic Energy Panel of the Defense Research and Development Board of the Department of Defense had held extensive discussions on this subject. Both, in fact, had made reports advocating accelerated research into the possibility of tactical nuclear weapons. Now the demands of the Korean War and the threat of general war made action imperative.

The matter of the technical feasibility of such smaller yield nuclear weapons was no longer in question: already feasible designs for such weapons existed. The issue was whether the nation should invest its scarce nuclear resources in them. With this question in mind, a number of scientists such as Charles Lauritsen of the California Institute of Technology and Louis Ridenour, Chief Scientist, U.S. Air Force, prevailed upon the military to establish a summer study project on the tactical use of nuclear weapons as well as some projects in other areas of national defense. Such summer studies undertaken in specific areas of combat were expected to substitute in part for the function which had been performed during World War II by the then defunct Office of Scientific Research and Development.

While the only summer study project of major concern to this history is Project Vista, a number of others should also be noted.[2] In addition to Vista there were two other major projects, Project Hartwell and Project Charles. Hartwell was devoted to the problem of undersea warfare and Charles to the problem of continental air defense. Charles gave rise to Project Lincoln (Lincoln Laboratory of the Massachusetts Institute of Technology) whose responsibility would be to apply modern electronic systems to the problem of air defense.

[2] See Huntington, *op.cit.*, for a discussion of these projects.

In this connection mention should also be made of Projects Summer Lincoln and East River which, respectively, explored further the matters of active and passive air defense. The former has some importance for this history in that its recommendation to the President that the United States construct a Distant Early Warning line (DEW line) in Northern Canada was to figure in Air Force criticisms of certain scientists and in the security trial of Robert Oppenheimer. Project East River proposed a strong civil defense program including urban decentralization and improved systems of warning of attack.

In a very real sense the purpose, or at least the result, of Project Vista, which was conducted on the campus of the California Institute of Technology in the summer of 1951 and completed in mid 1952, was to spell out in detail some of the notions of the finite containment scientists embodied in the GAC report on the hydrogen bomb. Even though members of the infinite containment school and some military officers participated in the project, the initiative behind the project was provided by finite containment scientists under the leadership of the GAC scientists. In fact the head of Project Vista was Lee DuBridge, the President of California Institute of Technology, who had been a member of the GAC at the time of the hydrogen bomb controversy. Other participants in the project included the following: Robert Oppenheimer, Jerrold Zacharias, Charles Lauritsen, Robert Bacher, Robert Christie, E. O. Lawrence, Walter Whitman, and John Fowler.[3]

The fundamental thesis underlying Project Vista was a rejection of the one weapon strategic concept, i.e., atomic bombs delivered by long-range big bombers, which dominated American military planning.[4] In seeking to counter this con-

---

[3] The total membership of the project is still held secret.

[4] The following discussion is based on the transcript of the Oppenheimer security trial, an article by Hanson Baldwin in *The New York Times*, June 5, 1952, p. 13, and a number of general references.

cept, Vista was to become the most thorough study of the strategic problems of the West as of that date (perhaps as of any date) and was to have enormous repercussions for Western defense although it was not to succeed in replacing the one weapon strategic concept.

As the previous chapter has indicated, scientists had sought for a number of years to convince the military services of the *tactical* value of atomic weapons and, at the same time, of the danger of an overdependence upon nuclear weapons. While the report of the GAC on the hydrogen bomb contained the most noteworthy of these attempts, it was but one of many made by those scientists on the GAC and on other advisory bodies who had played the major scientific role in the development of atomic weapons prior to the establishment of Project Vista.

Despite the fact that the United States was engaged in a limited war in which it was very unlikely that strategic air power would be used, three obstacles to a de-emphasis upon strategic bombing still existed. In the first place the psychology was still predominant in the military and, above all, in the Air Force, that there was a critical shortage of fissionable material and that consequently nuclear weapons must be used solely for strategic bombing and in particular for Air Force rather than Navy strategic bombing.[5] A number of leading Air Force officers did not believe there was sufficient nuclear material to warrant its diversion into relatively costly tactical weapons.[6]

A second obstacle to the exploitation of tactical nuclear weapons was the lack of sufficient tactical air power and of doctrine for the tactical use of atomic weapons. This matter essentially involved a dispute between the Army and the Air Force which had existed since World War II and had become

---

[5] This, it will be remembered, was at issue in the B-36 vs. super carrier controversy of 1949.

[6] General Curtis LeMay's belief that all fissionable material by right belonged to the Strategic Air Command because, in his view, there was no such thing as enough is discussed in James R. Shepley and Clay Blair, Jr., *The Hydrogen Bomb,* London: Jarrolds Ltd., 1954, p. 177.

acute following the military "unification" act of 1947. Whereas the Army desired a greater emphasis on tactical air power and advocated a tactical doctrine which would give it operational control over aircraft allocated to close ground support, the Air Force took the opposite view. The dominant view in the Air Force was one which de-emphasized tactical air power in favor of strategic bombing and which regarded air power as an autonomous mission not subordinate to the Army's tactical requirements.

The third obstacle facing the scientists in Project Vista was the belief that the United States economy could not afford increased expenditures on conventional armament. The Communist strategy, so this argument went, was to "bleed America to death." Paradoxically, this argument against the scientists' desire for a balanced military power would carry the day for the scientists' other argument that the United States ought to increase its tactical nuclear capabilities. The cost, however, would be a further decline in Western conventional capabilities.

By the time of Project Vista the scientists had done much to educate the military and the Administration to the potential value of tactical nuclear weapons. Among the many scientists who had sought to convert the military to a new strategic concept the most important had been Robert Oppenheimer.[7] Walter Whitman, former chairman of the Research and Development Board (RDB) of the Department of Defense, told the Gray Board in 1954, " . . . always Dr. Oppenheimer was trying to point out the wide variety of military uses for the bomb, the small bomb as well as the large bomb. He was doing it in a climate where many folks felt that only strategic bombing was a field for the atomic weapon."[8] Whitman credits Oppenheimer with doing "more

[7] In addition to personal qualities Oppenheimer's influence can be explained by his position as chairman of both the GAC and the Atomic Energy Panel of the Research and Development Board.

[8] U.S. Atomic Energy Commission, *In the Matter of J. Robert Oppenheimer,* U.S.G.P.O., 1954, p. 497.

than any other man . . . to educate the military to the potentialities of the atomic weapon for other than strategic bombing purposes; its use possibly in tactical situations or in bombing 500 miles back . . . . The idea of a range of weapons suitable for a multiplicity of military purposes was a key to the campaign which he felt should be pressed and with which I agreed."[9]

Prior to the convening of Project Vista, Robert Oppenheimer had asked the question which was to engage the attention of the Project members: "What contribution may one reasonably hope that the atom can make to our military power, the power for the prevention of war, the limitation of war, and for the defeat of the enemy in the event that war does come?"[10] Specifically, the responsibility of the scientists in Project Vista was to answer this question with respect to the defense of Europe, then threatened by the growing possibility of a general attack on the West by the Soviet Union.

The answer given by Project Vista to Oppenheimer's question was one which placed the scientists clearly on the side of the Army and against the Air Force in the dispute over national strategy. Robert Oppenheimer himself had foreshadowed this possibility when he had written in February 1951: "It is clear that they [atomic weapons] can be used only as adjuncts in a military campaign which has some other components, and whose purpose is a military victory. They are not primarily weapons of totality or terror, but weapons used to give combat forces help that they would otherwise lack. They are an integral part of military operations."[11]

The greatest emphasis in the report of Project Vista was on the need for an increase in American tactical air power and for a de-emphasis on strategic air power. Only under certain circumstances such as a direct full scale attack upon

---

[9] *ibid.*
[10] J. Robert Oppenheimer, "Comment on the Military Value of the Atom," *Bull. Atom. Sci.*, Vol. 7, No. 2, February 1951, p. 43.
[11] *ibid.*, p. 44.

Western Europe or the United States, the project report argued, ought the United States to employ the Strategic Air Command.¹² Instead the primary response to Communist limited aggression ought to be with conventional and tactical nuclear weapons.

While denying that conventional arms could be replaced by tactical atomic weapons, Project Vista did advocate that the latter be developed for use against troops, airfields, and supply lines. In order to carry this policy out the project further advocated two policies which were greatly to disturb the Air Force. Firstly, the report of Project Vista advised that the United States make a tripartite division of its fissionable materials among the three military services; tactical weapons would then be made available to the Army; the Navy would have both tactical and strategic nuclear weapons.¹³ Secondly, the project developed in elaborate detail a doctrine for the tactical use of nuclear weapons desired by the Army.

The participants in Project Vista believed that the tactical use of atomic weapons would prevent the Communist bloc from being able to mass its forces in secret behind the Iron Curtain in such a way as to be able to attack Western Europe with sufficient momentum to push the forces of the North Atlantic Treaty Organization (NATO) into the English Channel. Atomic weapons employed in land combat would therefore compensate for the alleged Soviet advantages of secrecy, of depth, and of superior numbers. The Russians would be forced to divide their forces in order to lessen their attractiveness as a target for atomic weapons, and this would

---

¹² The suggestion that the United States ought to announce this restriction on the use of SAC was in an earlier version of the report and may have been in the final report as well. The intent of this idea was to eliminate the possibility of the accidental expansion of limited nuclear war into total nuclear war.

¹³ Hanson Baldwin, *The New York Times,* June 5, 1952, p. 13, says that the project also sided with the Navy against the Air Force in the former's desire for a super carrier.

enable the lesser forces of NATO to deal with Soviet dispersed forces by use of conventional weapons.

Project Vista members argued that the effect of this change in tactics would be to increase the Western containment power in three ways. Firstly, in the same way that the strategic use of atomic weapons deters total war, their tactical use would deter limited war. The Soviet Union would be deprived of the hope of victory in Europe by either type of war. Secondly, this deterrence on the tactical level as well as on the strategic level would eliminate a potent propaganda theme of the Soviet Union. The USSR would no longer be able to taunt the Europeans with the idea that their security depended upon its good will because the United States could not be depended upon to defend them when America's only recourse in the face of Soviet aggression was a total atomic war in which it too would suffer great loss. Thirdly, and most importantly, an increase in the Western capability for a flexible response to Communist threats which would permit the renunciation of the use of strategic air power except under certain circumstances would strengthen the will of the West to resist limited Soviet aggression.

The emphasis in Project Vista reflected the particular period of tension in which it took place. The Chinese intervention in Korea had caused great fears that a general war for which the United States was quite unprepared would break out immediately. The threat of the A-bomb had failed to prevent the war in Korea, the possibilities for an imminent creation of a thermonuclear bomb which most of these scientists now desired were dim,[14] and the United States had insufficient conventional power to counter the Soviet initiative.

In making their policy recommendations, however, the scientists had added fuel to the inter-service argument over "roles and missions," i.e., which service would control the atomic bomb. At the Key West Conference in 1948 the issue had been decided in favor of the Air Force although the Navy was "not to be denied the use of the A-bomb"

[14] Ralph Lapp, *Atoms and People,* Harpers, 1956, p. 110.

against specific targets associated with its missions.[15] Now the scientists not only argued for a greater role for the Navy but they advocated full partnership for the Army. In addition, at a time when the Administration and Congress were desirous of leveling off and eventually decreasing the funds for conventional arms, the scientists advocated a greater emphasis on conventional armament. Lastly, the scientists had entered into the intra-Air Force debate over tactical vs. strategic aircraft which had been stirred up by the Korean War.

As a consequence of this forthright stand against elements in the Air Force and within both political parties the leading scientists in Project Vista earned the enmity of many influential persons. In part these feelings of hostility would account for the security trial of Robert Oppenheimer in 1954. They would also continue to the present day to create in many of these same persons a suspicion and distrust of the scientists who hold the finite containment point of view.

The American Hydrogen Bomb

While the scientists in Project Vista and others were seeking a viable alternative to overdependence on the doctrine of strategic air power, still other scientists such as Edward Teller, Hans Bethe, and John von Neumann were moving toward a reinforcement of that doctrine through their progress on the problem of the thermonuclear weapon. Whether their motivation was to prove its impossibility (Bethe), to open up the possibility of a new field of tactical weapons (Teller), or primarily to preserve the strategic balance (von Neumann), the final outcome of this research would lead to the enthronement of strategic air power as America's primary response to Communist aggression.

By 1951 it had become obvious that Teller's original idea

---

[15] See Walter Millis, *Arms and Men*, A Mentor Book, 1956, pp. 285–86. The suggestion in the Vista report that SAC be placed under the operational control of theater commanders such as Army Generals Dwight Eisenhower or Matthew Ridgway in the European theater was not calculated to please the Air Force either.

on the thermonuclear weapon known as the Super would not succeed; problems of temperature and material appeared to be insurmountable. However, in *Operation Greenhouse* in May 1951 a way was conceived by which the temperature problem could be solved. Most significantly of all, in December of 1950, Teller had thought of a new theoretical approach which would solve the material problem as well.

In June 1951 the principal nuclear scientists gathered at Princeton to hear and review Teller's new ideas. As the scientists who had originally opposed the bomb listened they realized that, as Bethe has put it, the hydrogen bomb had become "inevitable." It had, in the words of Oppenheimer, become "technically so sweet that you could not argue about that. It was purely the military, the political and the humane problem of what you were going to do about it once you had it."[16]

On the basis of Teller's new approach[17] and with the important additional aid of MANIAC, a high speed computer completed in 1951, the scientists set to work and, within eighteen months, had developed an experimental thermonuclear "device." This device, code-named MIKE, was exploded with the force of several million tons of TNT[18] at Eniwetok Atoll on November 1, 1952 as part of *Operation Ivy*.

Finally in March 1954—seven months after the first Russian hydrogen explosion—the United States achieved its first "droppable" hydrogen bomb; test shot BRAVO was exploded with a force equivalent to 15 megatons of TNT. The world had now entered the era of Megadeath and thus an unprecedented problem for American nuclear policy had appeared.

[16] U.S. Atomic Energy Commission, *In the Matter of J. Robert Oppenheimer, op.cit.,* p. 251.

[17] It is quite wrong to give Teller, as some do, complete credit or blame for the American thermonuclear weapon. For a discussion of this point see Edward Teller, "The Work of Many People," *Science,* Vol. 121, No. 3139, February 25, 1955, p. 275.

[18] Ralph Lapp, *Atoms and People, op.cit.,* p. 112.

The Defense Debate of 1953–1954:
The Doctrine of Massive Retaliation

THE SEARCH FOR A NEW LOOK

Oppenheimer's hope that American political leadership would face up to the "military, political and humane problem" which the hydrogen bomb would cause was to be short-lived. When the new Eisenhower Administration came into power in January 1953, its commitment to the reduction of defense expenditures was too strong to enable it to give appropriate thought either to this problem or to the dangers of overdependence upon nuclear arms raised by Project Vista. Furthermore, the success of the hydrogen bomb project gave the advocates of strategic air power the capstone to their argument for its invincibility.

Conservatives within the Republican Party, like Robert Taft and George Humphrey, demanded that the Republican pledge to reduce expenditures be central to the thinking of the new Administration. Therefore, when in the spring of 1953 the Eisenhower Administration took its first step to introduce a New Look in American military policy, the budget of every military service was reduced. Within the restricted limits of the new budget, emphasis was placed on atomic weapons, strategic bombing, and continental air defense. Although the United States had just emerged from a conventional limited war, these new policies could only reduce its capacity to fight another war of this type. As Secretary of the Treasury George Humphrey put it, the United States had " . . . no business getting into little wars. If a situation comes up where our interests justify intervention let's intervene decisively with all we have got or stay out."[19]

In their development of the New Look, the Republicans

[19] Quoted in Norman Graebner, *The New Isolationism,* Ronald Press Company, 1956, p. 132.

drew upon what they considered to be the principal lesson of the Korean War. They believed that the Truman Administration had made a serious mistake in implying that Korea was outside the zone of American nuclear retaliation. This mistake, in their view, had led to the Korean War, and the Republicans were determined not to make this same mistake again. Paradoxically they believed that the scientists in Project Vista had provided them with the means to contain the USSR at a tolerable cost through a primary emphasis on air power and atomic weapons.

In a Senate speech in September 1951 Senator Brien McMahon, a Democrat and chairman of the Joint Committee on Atomic Energy (JCAE), had already drawn attention to a solution to the problem of how defense expenditures could be reduced without sacrifice of military power. McMahon had opened his speech with the dire warning that the then "soaring defense costs" would lead to "national bankruptcy." Then, he hopefully said, "a coming revolution in military firepower points the way out. It points to a revolution in deterring power. It can bring peace power at bearable costs."[20]

Other senators, AEC Chairman Gordon Dean, and Secretary of Defense Robert Lovett, among others, endorsed Senator McMahon's sentiments. "Fortunately," said Secretary Lovett at a press conference the next day, "there is enough truth in both the weapons development stories and in the progress reports on atomic energy to encourage a very optimistic outlook for improved American armaments."[21]

The increased pace and nature of nuclear testing tells the story of this progress. While there had been only two tests (Bikini) between 1945 and 1948 and four test shots (*Operation Sandstone*) in 1948, there were twelve low yield tests in 1951 (*Greenhouse*) and about the same number in

[20] Brien McMahon, "Atomic Weapons and Defense," *Bull. Atom. Sci.*, Vol. 7, No. 10, October 1951, p. 297.
[21] Quoted in "A News Chronology—Weeks of History and Mystery," *Bull. Atom. Sci.*, Vol. 7, No. 10, October 1951, p. 303.

1952.[22] Significantly, following the establishment of the Nevada Proving Grounds and of the Sandia Corporation, the AEC was rapidly developing small yield weapons (less than thirty kilotons). Furthermore, tactical nuclear weapons had captured the imagination of political leadership as the solution to the problem of limited Communist aggression.

THE DISARMAMENT PANEL'S RECOMMENDATIONS

Again Robert Oppenheimer was to find himself in the middle of a major controversy over national defense. While he was no longer on the General Advisory Committee to the Atomic Energy Commission, he had been appointed in the last days of the Truman Administration as chairman of a special State Department Advisory Committee on Disarmament whose membership also included Vannevar Bush, Allen Dulles, John Dickey, and Joseph Johnson. In July 1953 he published an article in which, by use of the most guarded and cautious language, he sought to convey to the public the program which his committee had put before the new President. The main purpose of the article was to lay bare the implications of the fact that the United States had made itself totally dependent militarily upon nuclear weapons and in particular upon the strategic employment of such weapons. "It is the only military instrument which brings the Soviet Union and the United States into contact—a most uncomfortable and dangerous contact—with one another."[23]

Oppenheimer went on to warn his countrymen that there were three things they must never forget: "One is the hostility and power of the Soviet. Another is the touch of weakness—the need for unity, the need for some stability, the need for armed strength on the part of our friends of the free world. And the third is the increasing peril of the atom." And, most of all, he added, "We need the greatest attainable freedom

[22] Ralph Lapp, *Atoms and People, op.cit.,* pp. 97–102.
[23] J. Robert Oppenheimer, "Atomic Weapons and American Policy," *Bull. Atom. Sci.,* Vol. 9, No. 6, July 1953, p. 203.

of action;" that is, we needed the freedom of action necessary for negotiation.[24]

Oppenheimer and the Disarmament Panel pressed upon the Eisenhower Administration three reforms in particular which they believed would contribute greatly to achievement of the unity, stability, and flexibility needed:

1. The United States had to give the American people the facts on the nature, implications, and requirements of the nuclear arms race in order to "make available to itself our inherent resources." The Administration ought to speak *with candor* to the people about the need for adequate defense. The Truman Administration had left the people ignorant through secrecy even though this secrecy was a cause of weakness because it permitted apathy. The Disarmament Panel believed the government had to educate the public to a sense of urgency if the nation were to build a defense adequate to meet Soviet advances.

2. The second recommendation was to increase Western unity in the face of the Soviet threat through greater candor with America's European allies. The first step in this direction would be to include them more in policy formation. Furthermore, these allies would have to have a greater voice in their own defense if they were not to succumb to Soviet threats. Specifically, the United States ought to increase their ability to defend themselves by sharing atomic weapons information with them.

3. The third recommendation was that development of continental air defense should be speeded. There were many opportunities for developments which could reduce America's vulnerability and thus decrease the Soviet incentive to attack. While the ultimate defense would be disarmament, an effective continental air defense would enhance American freedom to negotiate a disarmament agreement.

[24] *ibid.*, p. 204.

THE FATE OF THE PANEL'S RECOMMENDATIONS[25]

While the recommendations of the Disarmament Panel awaited disposition by the President, for the second time in a decade a Soviet nuclear explosion burst in upon the American deliberations. On August 12, 1953, the United States detected the explosion of a Soviet hydrogen bomb, and American policy was thrown into a state of flux and confusion. The rapid and unanticipated rate of Soviet technological achievement in addition to Soviet conventional power placed the evolving New Look in jeopardy. An administration dedicated to the reduction of military expenditures was faced with a military-technological challenge greater than that yet faced at any time in the American past. The Administration's disposition of the Disarmament Panel's recommendations reflected this ambivalence.

The need for candor. Robert Donovan, in his book *Eisenhower, the Inside Story,*[26] traces the history of the transformation of Operation Candor as proposed by the Disarmament Panel into the Operation Wheaties carried out by the Administration. In April 1953 President Eisenhower decided that he should speak to the public on aspects of the nuclear arms race. He would inform the nation of the significance of the Soviet hydrogen explosion and of the nature of the Soviet challenge. However, as the President prepared his speech, both external and internal pressures were being exerted to convince him that he should present the challenge of Soviet technological progress in atomic weapons with optimism rather than with pessimism. His task, it was argued, was to give Americans a ray of hope. Of decisive importance, however, was the view that a frank statement of America's weaknesses from the

---

[25] The disposition of the panel's recommendation on continental defense is a subject in itself and outside the scope of this study; for a discussion of this subject, see Huntington, *op.cit.,* pp. 326ff.

[26] Harpers, 1956, pp. 184–86.

President would have a detrimental effect on its international position.[27]

After much tortuous weeding out of speeches and proposals, Lewis Strauss, Chairman of the Atomic Energy Commission, struck the note the President was seeking. Strauss' Operation Wheaties proposal was mid-range between the pessimistic position taken by the Disarmament Panel and the approach suggested by the Chairman of the Joint Chiefs of Staff that the President inform the American people they were secure because the United States was superior to the Soviets in its atomic stockpile.

The solution found in Operation Wheaties was for the President to touch lightly in his speech on the horrors of the hydrogen bomb, to play up America's "peace" power, and to conclude with the now famous "Atoms for Peace" proposal. This "new imaginative approach" to disarmament would supposedly attack the problem indirectly. The nations would establish an International Atomic Energy Agency to which the atomic powers would each turn over a small part of their stockpile of fissionable materials. This literal conversion of "swords into plowshares" would symbolize the great benefits to man of international cooperation in the atomic field. Hopefully a new atmosphere of international cooperation surrounding the atom would then be created and one day a genuine solution to the threat of the hydrogen bomb would come out of this atmosphere. Ironically, the main effect of the Atoms for Peace Plan, as Edward Teller has since pointed out, may be to accelerate the capability of many nations to produce their own nuclear weapons.

Nuclear sharing with America's European allies. The Disarmament Panel had argued that America's European allies would need an atomic capability of their own if the West were to meet the alleged Soviet strategic advantages of manpower, geography, and secrecy after the arrival of nuclear parity. This

[27] See Robert Cutler, "The Seamless Web," *Harvard Alumni Bulletin,* June 4, 1955, p. 665.

required the exchange of nuclear weapons information to enable Europeans both to employ tactical atomic weapons and to prepare defenses against atomic weapons.

In his State of the Union message, delivered January 7, 1954, President Eisenhower indicated his support for this idea; the President proposed an amendment to the Atomic Energy Act of 1946 which would permit him to supply information on atomic weapons to America's European allies. The resultant legislation, however, was too restrictive to create the tactical atomic weapons capability in Western Europe desired by the President. Nonetheless, on December 17, 1954, the NATO Council decided to follow the recommendation of the Disarmament Panel (and of Project Vista as well) that the ground defense of Europe should employ tactical nuclear weapons.

THE DOCTRINE OF MASSIVE RETALIATION

Secretary of State John Foster Dulles had entered office with the conviction that Secretary Acheson had erred when he had defined the threshold of American nuclear retaliation and that as a consequence the United States had had to fight the bloody, costly Korean War. Such wars, in Dulles' view, were to the advantage of the Communists with their willingness to expend resources and manpower in order to bankrupt the United States.

Dulles believed that the threat of atomic retaliation could prevent such wars of attrition only if the threat were an ambiguous one. The United States must not "call its shots" ahead of time as Acheson had done with respect to Korea; the potential aggressor must be kept in ignorance of America's intentions. Dulles enunciated this new American policy of massive retaliation in the following words: "A potential aggressor must know that he cannot always prescribe the battle conditions that suit him . . . . The basic decision . . . [is] to depend primarily upon a great capacity to retaliate instantly by means and at places of our choosing. And now the Department of Defense and the Joint Chiefs of Staff can shape our military establishment to fit what is our policy instead of having to try to be ready to meet the enemy's many choices. And that permits a selection of military means instead of a multiplication of means. As a result it is now possible to get, and to share, more basic security at less cost."[28]

In contrast to the scientists' view which had been recently expressed in the Project Vista recommendations that the function of American strategic air power was only to deter Soviet strategic air power, Dulles proposed that the threat of American strategic air power be used to deter *all* types of Communist aggression. Whereas the scientists preferred that local aggression be met by local defense, Dulles actually wanted to rule out wars of attrition. In the words of Vice President Richard Nixon, "rather than let the Communists nibble us to death all over the world in little wars, we would rely in the future primarily on our massive mobile retaliatory power . . . against the major source of aggression . . . ."[29]

In this capsule version of Secretary Dulles' new doctrine, the Vice President confirmed the fact that the new Administration had endorsed a military policy against which scientists had fought for almost ten years. With few exceptions scientists of all persuasions rejected the new policy. Their attempts to overturn it will absorb the rest of our history.

Conclusion

The swift movement of events after the outbreak of the Korean War left the scientists in a state of confusion. This was especially true of the finite containment scientists who had supported the development of tactical nuclear weapons as a means to provide a realistic alternative to strategic air power. Slowly the finite containment scientists were to realize that

---

[28] *The New York Times,* January 13, 1954, p. 2. Dulles' doctrine was enunciated with the then current Indochina War in mind. The United States was seriously considering intervention and at the least hoped that a threat of intervention could save the hard-pressed French.

[29] Quoted from Radio Address of Vice-President Richard Nixon, March 13, 1954, in the *Bull. Atom. Sci.,* Vol. 10, No. 5, May 1954, p. 149.

they had contributed to a worsening of the problem of atomic energy—at least in the terms within which they themselves viewed the problem. On the one hand, these tactical nuclear weapons did appear to be necessary, given the weaknesses in American conventional capabilities; the only alternative seemed to be total dependence upon strategic bombing. However, on the other hand, if the nations should replace conventional armaments with tactical atomic weapons, they might then never be able to disarm atomically.

This ambiguity with respect to tactical atomic weapons was well expressed by David Inglis as early as 1952: "Thus the short-range considerations of the defense of Europe, and perhaps also of South Korea, seem to dictate immediate emphasis on the tactical use of atomic weapons, and the short-range prospects appear to be brightened by the possibility of placing at least partial reliance on these weapons."[30] Yet Inglis went on to point out that "in the long-range view . . . the news of tactical . . . atomic explosions is not good news."[31] He argued that they really were not substitutes for conventional weapons; in his view there would always remain a qualitative difference between nuclear and conventional explosives because there was no continuum in explosive power between them.[32] Yet Inglis believed that there was still time to negotiate tactical nuclear weapons out of existence. Inglis saw little hope, if this were not done, for the eventual success of the international control of atomic energy.

The man who had done the most to bring about this interest in tactical nuclear weapons even while also warning against overdependence upon them would soon be dealt a grave injustice in part because of his efforts. Many influential persons did not regard the activities of Robert Oppenheimer in a

[30] David Inglis, "Tactical Atomic Weapons and the Problem of Ultimate Control," *Bull. Atom. Sci.,* Vol. 8, No. 3, March 1952, p. 79.
[31] *ibid.*
[32] This belief, of course, has proven to be incorrect. Some nuclear weapons, such as the Davy Crockett, have an explosive force less than some conventional weapons.

kindly light. In particular, certain Air Force representatives were concerned over his behavior. "It became apparent to us," David Griggs, Chief Scientist for the Air Force, told the Gray Board, " . . . that there was a pattern of activities all of which involved Dr. Oppenheimer. Of these one was the Vista project. . . . It was further told me . . . that in order to achieve world peace . . . [Oppenheimer believed] it was necessary . . . to give up . . . the strategic part of our total air power."[33]

A few months before these words were spoken, the train of events had begun which was soon to eliminate the valuable talents of Robert Oppenheimer from governmental service. In the fall of 1953, William L. Borden, a dedicated believer in the theory of strategic air power[34] and a former executive director of the staff of the Joint Committee on Atomic Energy, had written a letter to FBI Director J. Edgar Hoover. In that letter Borden had stated his "own exhaustively considered opinion, based upon years of study, of the available classified evidence, that more probably than not J. Robert Oppenheimer is an agent of the Soviet Union."[35]

The subsequent security trial, which resulted in the denial of security clearance to Oppenheimer and consequently in his disbarment from governmental service, brought to an end the public career of a dedicated patriot who had served his country well. He had made valuable contributions to the atomic bomb, Project Vista, *NSC68*, the hydrogen bomb, and many other projects. His alleged failure to be "candid," his alleged "disregard for the requirements of the security system,"

---

[33] U.S. Atomic Energy Commission, *In the Matter of J. Robert Oppenheimer, op.cit.*, p. 749.

[34] For his views see William L. Borden, *There Will Be No Time*, Macmillan, 1946. This is an excellent exposition of the doctrine of neo-isolationism. It argued (1) that American foreign policy should be based on strategic air power and (2) that the major threat to this power is internal espionage.

[35] U.S. Atomic Energy Commission, *In the Matter of J. Robert Oppenheimer, op.cit.*, p. 837.

and his apparent indiscretions, all of which were believed to disqualify him on legal grounds for further governmental service, are points of controversy outside the scope of this study. His contributions to American security, the impact of his trial upon scientists, and the regrettable manner in which he was treated by the government are, however, significant for this history of the participation of scientists in the formulation of American nuclear policy.

In a very real sense, the Oppenheimer security trial meant the end of innocence for the scientists. It also brought to an end another period in the political struggles among scientists and bitterly accentuated the political breach within the scientific community. And with the Administration's implication that Oppenheimer could not be trusted because of his opposition to the hydrogen bomb, the government estranged many scientists and weakened their faith in the competence and integrity of American political leadership.

For the first time in almost a decade of political activities, the restraining code of professional ethics which had governed scientists' actions was nearly completely submerged by latent passions of resentment and hostility, and scientists committed the "unpardonable" sin of questioning one another's personal integrity. While the emotional attacks were directed primarily against Oppenheimer, other scientists were not left unspared.

A great many scientists were particularly disturbed over what seemed to be a personal attack by Teller upon Oppenheimer. Although Teller stated in response to questioning that he believed Oppenheimer to be a loyal American, he then was questioned further and replied as follows:

Question by Government Counsel: "Do you or do you not believe that Dr. Oppenheimer is a security risk?"

Answer: "In a great number of cases . . . [he has acted] in a way which for me was exceedingly hard to understand. I thoroughly disagreed with him in numerous issues and his actions frankly appeared to me confused and complicated.

To this extent I feel that I would like to see the vital interests of this country in hands which I understand better, and therefore trust more."[36]

The scientific community has yet to recover from the traumatic experience of the Oppenheimer security trial. Not only did the trial aggravate already existing distrust and political disagreement but it hardened this cleavage by the addition of serious personal and organizational antagonisms. Scientists not only chose to align with either Teller or Oppenheimer, but the split between the Los Alamos Laboratory and the Livermore Laboratory became even more definite.[37] The trial thereby accentuated the struggle for power among certain scientists in a manner not always conducive to the national interest.[38]

[36] *ibid.*, p. 710.

[37] The Livermore Radiation Laboratory was founded in 1952 by the AEC in response to pressures by the Air Force and Edward Teller who was dissatisfied with Los Alamos. While Livermore contributed little to the initial development of the hydrogen bomb, it subsequently made many contributions to the weapons program of the United States.

[38] For a discussion of the significance of the Oppenheimer case in the eyes of many scientists see the following three issues of the *Bull. Atom. Sci.*, Vol. 10, No. 5, May 1954; Vol. 10, No. 6, June 1954; and Vol. 10, No. 7, September 1954. A balanced treatment of the case is to be found in J. Stefan Dupré and Sanford A. Lakoff, *Science and the Nation—Policy and Politics,* Prentice-Hall, 1962, pp. 141–50.

CHAPTER FIVE

THE EMERGENCE OF THE "FIRST STEP"
PHILOSOPHY

THE politics of the scientists over the issue of nuclear weapons has been divided into several distinctive stages. The first period from 1945 to late 1947 was one in which the scientists were relatively unanimous concerning political ends and means. The second period from 1947 to 1949 was one which began with confusion of purpose, developed a split between the control and containment schools, and ended in extreme antagonism between the finite and infinite containment schools. The third period from 1950 to 1955 was again one of relative unanimity as far as policy was concerned, although the hostility aroused among scientists by the hydrogen bomb question and the Oppenheimer security hearings accentuated the conflicts among the scientists and between them and the government. The post-1955 period which extends into 1962 and to which the remainder of this study will be devoted is closest in character to the 1947–1949 period. There is great confusion among scientists accompanied by cleavage and hostility within the scientific community.

In the period from 1947–1949, the political cleavage among scientists was generated by the general question of whether it was wiser to risk falling behind in the arms race in order to concentrate on reaching an agreement on a control system or to forego attempts to do this until political changes in the world decreased the risk implicit in such a course. This issue culminated in the latter part of this period in the question of whether or not to increase the magnitude of available destructive power through the construction of the hydrogen bomb.

In the period from 1955 through 1962, the primary issue has remained that of international control versus containment by nuclear weapons. However, the specific problem on which recent conflict has focused is whether or not to continue the development and improvement of specialized nuclear weapons for offensive and defensive purposes such as tactical, anti-missile, and strategic nuclear weapons. On one side are those scientists who believe that the United States must continue to develop new generations of nuclear weapons and to base its defense primarily on tactical as well as large strategic nuclear weapons. On the other side are the scientists who argue that such increases in national power would be negated by the enhanced threat to civilization caused by the nuclear arms race.

In face of the need to make new decisions on nuclear weapons United States policy has been in a transitional period since 1955 similar to that prior to the hydrogen bomb decision. The question to be decided has been whether to end the nuclear arms race or to proceed with the development of new defensive and offensive nuclear capabilities. The necessity for such a decision has given rise to the most passionate politics among scientists since World War II. Not only has the conflict between the two containment schools reappeared but a very vociferous control school has re-emerged and has joined forces with the finite containment school in opposition to the infinite containment school.

The divergence of the two containment schools upon the desirability of the continued development of specialized nuclear weapons is caused primarily by their different reactions to the convergence of a number of changes in the global environment. These changes have reinvigorated the view of the finite containment scientists which had been dormant since the beginning of the Cold War that international control of atomic weapons is imperative; in contrast, the infinite containment scientists have remained convinced that only the existence of an open world could make international control a safe policy for the United States.

By 1955 a new context of international politics was evolving in which a new attempt to achieve international control began to seem far more important to the finite containment scientists than did an increase in the nuclear capabilities of the United States. In this belief they were joined by the control school which was especially concerned over the dangers of radioactive fallout. As a result there emerged a strong scientific movement, in part rationalistic in nature, which assumed a responsibility to convince the nation of the necessity to take a "first step" toward disarmament.

The New Context of the Politics of Scientists

It is easily perceived that the specific issues at stake in this latest stage of the nuclear arms race are even less well defined than they were in 1949. Then the issue was the comparatively simple one of whether or not to develop the hydrogen bomb. The issues involved in the questions of atomic energy control and nuclear disarmament faced since 1955 have been extremely complex and interwoven. These issues arise from the military, political, and technical changes which had become apparent by 1955.

The change in the character of the problem of the control of atomic weapons is due to a variety of factors which together seem to many people to increase both the desirability of and the hopes for the establishment of a system of international control. These factors are the following: (1) fear of radioactive fallout; (2) the Soviet "thaw"; (3) the advent of nuclear "parity" and mutual deterrence; (4) changing nature of the arms race; (5) the apparently decreasing technical feasibility of disarmament; and (6) the problem of the Nth power.

FEAR OF RADIOACTIVE FALLOUT

The failure of the Truman and Eisenhower Administrations to educate the American public to the complex and fearful nature of the atomic arms race was to plague the government sooner than might have been anticipated. Less than two months following the substitution of Operation Wheaties for the proposed Operation Candor, an event occurred which was to arouse storms of criticism of the atomic weapons program of the United States. On March 1, 1954, the United States tested a nuclear weapon of super proportions; test shot Bravo was equivalent to fifteen million tons of TNT and its cloud burst through the tropopause, spilling nuclear debris into the upper stratosphere.[1] The movement of the debris upwards and outwards was greater than had been expected by American military and scientific personnel, and the "ashy rain" fell a hundred miles from the bomb site onto a small Japanese tuna trawler named the *Lucky Dragon Number 5*.

Two weeks later the whole world had heard of the voyage of the *Lucky Dragon*. The fishermen were afflicted with radiation sickness, and allegedly from the indirect results of this one man was to die. However, the greatest impact came from the reported discovery that contaminated radioactive fish from the test area were reaching the Japanese market. Hysteria swept Japan; American-Japanese relations became strained and the whole world resounded with criticism of the United States.

At the initiative of Lewis Strauss the United States Government responded quickly with offers of aid which did much to ameliorate the situation; nevertheless, the Bravo explosion and its consequences had raised new and serious questions. Although the fallout from atomic weapons had been considered potentially dangerous as early as 1945, it had thus far been believed to be relatively inconsequential.[2] However, the evidence that now trickled in from Japan indicated that radioactive fallout had added a new dimension to be con-

[1] Jack Schubert and Ralph Lapp, *Radiation—What It Is and How It Affects You*, The Viking Press, 1958, p. 220.

[2] In the light of subsequent developments it is curious that one of the first, if not the very first, to suggest publicly that fallout constituted a serious danger for the human race had been Edward Teller. "How Dangerous Are Atomic Weapons?" *Bull. Atom. Sci.*, Vol. 3, No. 2, February 1947, pp. 35–36.

sidered in the problem of nuclear testing and atomic warfare.

In response to charges that the Administration had been irresponsible in carrying out the Bravo shot, that the explosion had gone out of control, and that the fallout from Bravo constituted an immediate health hazard, President Eisenhower held a news conference on March 31, 1954, which AEC Chairman Strauss attended. In addition to explaining the circumstances of the Bravo explosion and the causes of the regrettable *Lucky Dragon* incident, Strauss sought to allay fears concerning the danger of fallout. In essence, Strauss argued that the level of radioactivity in both short-range and long-range fallout would "decrease rapidly after the tests until the radiation level has returned approximately to the normal background."[3]

This explanation failed, however, to satisfy the critics of the Commission's testing policy and it was believed by many in the scientific community and others including members of the Joint Committee on Atomic Energy that the Atomic Energy Commission was deliberately withholding information from the public on this rather frightening development. These suspicions and a lack of confidence in the Atomic Energy Commission and in its chairman, Lewis Strauss, were intensified as the results of Japanese scientific studies on the radioactive residue of the Bravo shot were made available throughout 1954. The findings resulting from the chemical and physical analysis of fish convinced many scientists that the Atomic Energy Commission and its chairman were trying to hide the dangerous nature of the atomic arms race from the American people.

The Japanese data indicated to these scientists that the

[3] Press release of the U.S. Atomic Energy Commission, March 31, 1954, p. 4. The difference between local and long-range fallout is that the former is composed of heavy radioactive particles which fall near the explosion and the latter are smaller fission products which are scattered by the prevailing winds over the globe through the mid-latitudes. The large Bravo explosion produced great quantities of both. Prior to that explosion, the danger of long-range radioactive fallout had been believed generally insignificant.

hydrogen bomb exploded in the Bravo test had been encased in U238, the relatively cheap and plentiful isotope of uranium, and they realized that the destructiveness of such a fission-fusion-fission bomb[4] could be increased to fantastic proportions. It seemed apparent to many scientists that the United States was deliberately making its bombs "dirtier"; these were obviously weapons whose potency could destroy much of mankind. Furthermore, if the United States was making such weapons, surely the Russians, who were well aware of this fact through the Japanese studies if not their own, were doing the same.

The official Atomic Energy Commission release on the fallout of the Bravo test appeared in February 1955, eleven months after the explosion; it did little to allay the concern of scientists. In fact, many scientists viewed the report as a distortion intended to calm the fears of the nation. The report minimized the local fallout effect such as that which had contaminated the Japanese fishermen and claimed that the genetic effect of this fallout was expected to be nil. In addition the long-range pathological effects of world-wide fallout were indicated to be negligible. However, the report appeared to leave more questions unanswered than answered, particularly on the nature of the Bravo bomb itself and on the exact extent of its radioactivity.

A striking example of scientific reaction to the AEC report on Bravo was the charge made by Dr. George LeRoy, Associate Dean of the Biological Sciences, University of Chicago, that the Atomic Energy Commission was holding back information vital to medical civil defense. Strauss characterized these latter remarks as "irresponsible." As LeRoy had been in charge of biological medical research at the Eniwetok Proving Grounds, his scientific colleagues tended to support his views.

---

[4] The fission-fusion-fission bomb is otherwise known as the uranium bomb. Its inner core is a fission bomb (U235 or plutonium); its outer core is a fusion bomb (hydrogen); and its casing is uranium 238. The U238 is transmuted into radioactive materials along with the inner core of the bomb, thus making it a very "dirty" bomb.

The findings of the Japanese scientists which were publicized by Ralph Lapp[5] and other American scientists, in addition to the apparently secretive attitude of the Atomic Energy Commission thus resulted in a serious loss of confidence in the Administration by many members of the scientific community. Although the Eisenhower Administration had seemed to give backing to disarmament in the "Atoms for Peace" proposal and although there were signs of awareness of the dangers of imbalance in the American military forces, the evidence that the United States had created a fission-fusion-fission bomb indicated to many scientists that the government was actually moving in the dangerous opposite direction. Just as the original hydrogen bomb had been the response of the Truman Administration to the Soviet atomic bomb, this newest version appeared to be the Eisenhower Administration's response to the Soviet hydrogen bomb. Increasingly many scientists believed that American military policy was continuing down the path to the inevitable fork in the road where it would have no choice in the face of Communist aggression other than a surrender to the Soviets or mutual annihilation.

These fears concerning the course of American nuclear weapons policy were reinforced by advances in the science of genetics and by studies on the pathological effects of radioactivity. American biologists had been studying the effects of radioactive fallout from the point of view of its genetic and pathological damage for some time. In this post-1955 period these studies were beginning to produce a body of data which was quite disturbing. It had long been known that radioactive materials from atomic bomb tests were carried by winds around the globe where they settled down to earth in the middle latitudes. It had been assumed, however, that this long-range fallout was inconsequential, an assumption supported in part by the fact that study of the

---

[5] Ralph Lapp, "Civil Defense Faces New Peril," *Bull. Atom. Sci.,* Vol. 10, No. 9, November 1954, pp. 349–51 and "Fallout and Candor," *Bull. Atom. Sci.,* Vol. 11, No. 5, May 1955, p. 170.

Hiroshima population showed no effect of radiation other than the pathological effects of the bomb itself.

Although there had been some who argued on the basis of genetic theory that genetic damage would result from the radioactivity released by atomic weapons, there was no significant support for this conclusion until 1949.[6] Instead, it had been assumed that the genetic effect of irradiation was not proportional to the dosage received but dependent upon the achievement of a certain threshold of irradiation. This is to say that there would be no genetic mutations produced unless a certain level of irradiation were reached. However, new laboratory evidence indicated that the extent of the genetic damage was proportional to the amount of the irradiation received and that the effect of irradiation on genetic material was cumulative in nature.

Most significant of all, the fact that 95 per cent or more of all mutations are harmful meant to many biologists that, theoretically, even a small increase in the radioactivity level of the world was harmful to man.[7] Therefore, it was becoming increasingly obvious to the biologists that there was a high probability that the testing of atomic weapons in the atmosphere was inflicting a genetic burden on future generations of men. Warren Weaver has summarized the issue for the layman as follows: "radiation produces mutations; mutations are bad; the amount of mutation is proportional to the total accumulated dose. There is no radiation dose which is too small to count from the point of view of genetics."[8]

A second source of concern for many scientists was the discovery that certain of the radioactive isotopes contained

[6] See the article by Delta E. Uphoff and Curt Stern, "The Genetic Effects of Low Intensity Irradiation," *Science*, Vol. 109, No. 2842, June 17, 1949, pp. 609–10. For his own understanding of this matter the writer is indebted to Dr. Hardin Jones, Donner Laboratory, University of California at Berkeley.

[7] H. J. Muller, "The Genetic Damage Produced by Radiation," *Bull. Atom. Sci.*, Vol. 11, No. 6, June 1955, pp. 210–12.

[8] U.S. Senate, Committee on Foreign Relations, *Hearings on Control and Reduction of Armaments*, 85th Congress, 1st Session, 1957, p. 1138.

in the long-range fallout were extremely long-lived and had a particular affinity for human tissue. The most notable of these isotopes was strontium 90 with its half-life of 28 years. Studies indicated that strontium 90 had chemical properties close to calcium, and as a result of this, it was being absorbed by plants which were subsequently eaten by cattle. The milk of these cows which then contained strontium 90 was given to children whose growing bones have a strong affinity for calcium and consequently for the strontium 90. As the bone marrow produces the body's white blood corpuscles, the end result of this process was feared to be the danger of causing dreaded leukemia in these children.

The growing evidence that the arms race and nuclear tests were causing such harmful effects created a revulsion among an increasingly large number of scientists, especially among the biologists who were most vividly cognizant of the effects of the testing program while also least familiar with its purpose. The Administration was regarded as acting irresponsibly in the face of the known consequences of testing, and there was a falling away of support for the government's weapons program. The growing protest of the biologists and others against the Administration's testing policy and the apparent failure of the physicist-representatives on advisory committees to express the biologists' case to the Administration would soon erupt into a vocal demonstration against nuclear testing under the leadership of Linus Pauling.

THE SOVIET THAW

While these events were taking place, developments were under way within the Soviet Union which were to have a great impact on the attitudes generated by the Bravo explosion and the biological studies. In the Soviet Union, following the death of Stalin in 1953, a noticeable "thaw" was observed to be taking place. Among the changes were numerous signs that the Soviet Union was emerging from its apparent myopia concerning the effects of atomic weapons. On March 12, 1954,

Premier Georgi Malenkov had overthrown the Soviet dogma that a nuclear war would mean the end only of capitalist and not of Communist civilization. "The Soviet Government . . . is resolutely opposed to the policy of cold war, for this is a policy of preparation for fresh world carnage, which, with modern methods of warfare, *means the ruin of world civilization.*"[9]

These remarks were followed on May 10, 1955 by a "major concession" of the Soviet Union on disarmament. After years of negotiation within the United Nations which had commenced with the Baruch Plan, the Russians finally accepted the Western position on conventional and nuclear disarmament, or at least so it appeared. The new Soviet plan seemed to accept Western demands concerning inspection and control as well as the Western belief in the inter-relatedness of nuclear and conventional disarmament.[10] They also proposed that a system of ground control posts was the method by which surprise attack should be prevented, and they recognized the problem of accounting for nuclear stockpiles in any disarmament system. Even though the Soviet Union never elaborated on these "concessions" and placed three conditions on disarmament—elimination of overseas United States bases, cessation of nuclear weapons tests, and outlawing of atomic weapons—the world believed that a greater flexibility was entering Soviet policy.[11]

It seemed to many scientists that, at long last, their 1945 prediction was coming true. The Soviet Union no longer denied the "logic of the facts" that in the atomic age there was only one choice open to man: either to control atomic energy or to perish by it. It appeared that, after a decade

[9] Quoted from *Pravda*, March 13, 1954, in Henry Kissinger, *Nuclear Weapons and Foreign Policy,*" Harpers, 1957, p. 383. (Italics Kissinger's.) The status or meaning of this thesis in Communist doctrine remains in doubt.

[10] Philip Noel-Baker, *The Arms Race*, London: Stevens and Sons, 1958, p. 20.

[11] U.S. Department of State, *Disarmament—The Intensified Effort, 1955–58*, U.S.G.P.O., 1958, p. 14.

of intransigence, the Soviet Union was awakening to reality. Most scientists believed that the West had a responsibility to seize this opportunity and to work out an agreement on atomic energy with post-Stalinist Russia.

THE ADVENT OF NUCLEAR PARITY AND MUTUAL DETERRENCE

The third factor which increased the desirability and the hope of achieving international control of atomic energy was the arrival of nuclear parity and mutual deterrence. In the fall of 1953 the Soviet Union had exploded its own hydrogen bomb and, although the Soviet Union was probably still behind the United States in the overall size of its nuclear arsenal, by 1955 it possessed or would soon possess a capacity to destroy the United States. The achievement of this nuclear parity meant that a period of mutual deterrence was near. Thus many scientists believed that if the arms race could be arrested at this point, this "balance of terror" could ensure the maintenance of peace, at least for a period of transition to total nuclear disarmament.

Moreover, many scientists believed that nuclear parity would make the Soviet Union more reasonable. As has been noted previously, a large number of scientists had believed that the intractability and apparent irrationalism of the Soviet Union was due to the fact that its inadequacy in the field of atomic weapons made it try to balance a weakness in power through harshness in language and through acts of belligerency. Now, according to this position, the Russian advances in atomic technology would remove the inferiority complex and the insecurity under which the Soviet Union had labored for the length of its existence and it would become more amenable to reason.

Leo Szilard later expressed this belief that the arrival of nuclear parity provided real hope for world peace: "A few years ago, with the increasing accumulation of bombs in the stockpiles of Russia as well as America, and with the progressive development of the means of delivery, a new factor became operative, and there began an at first almost imperceptible wavering in the seemingly inexorable course of events. It is my contention that, as the world moves into the next stage, the vicious circle of classical power conflict will cease to operate between America and Russia."[12]

In the thinking of at least the finite containment scientists, the implications of nuclear parity for the strategy of limited nuclear war which these scientists had advocated in Project Vista were undoubtedly more influential. In 1951–52 these men had reasoned that a limited nuclear war based on the tactical employment of nuclear weapons was a viable strategy for the West because they had assumed that the Russians would be forced to invest all their scarce fissionable materials in strategic weapons; consequently, they believed it would be a long time before the Russians could develop tactical nuclear weapons and cancel the American advantage in this category of weapons. The rapid development of Russian capabilities in both strategic and tactical nuclear weapons eroded the confidence of these scientists in the argument that the development of tactical nuclear weapons gave the West a strategic advantage. Thus in a world where both East and West could employ such weapons these scientists reversed their position on the tactical value of nuclear weapons, and they began to argue instead that further development of nuclear weapons in any category was to the disadvantage of the Western nations.

CHANGING NATURE OF THE ARMS RACE

A further consequence of the arrival of nuclear parity was the shift in the nature of the arms race; this shift was to become an additional factor in the alteration of the international environment after 1955. The arrival of nuclear parity

---

[12] Leo Szilard, "How to Live with the Bomb and Survive," *Bull. Atom. Sci.*, Vol. 16, No. 2, February 1960, p. 61.

meant, as Robert Oppenheimer had predicted in 1953, that the time would soon come when " . . . the art of delivery and the art of defense will have a higher military relevance than the supremacy in the atomic munitions field itself."[13]

One development in particular was beginning to command the attention of scientific advisors to the government, a development which was an unanticipated consequence of the thermonuclear weapons program of the Atomic Energy Commission: the potential perfection of compact, high yield warheads deliverable by ballistic missiles. Whereas, for a variety of technical and economic reasons, many scientific advisors had previously regarded the idea of an intercontinental ballistic missile (ICBM) as an absurdity, this development in nuclear warheads plus other advances in rocketry and electronics appeared to make an ICBM a virtual certainty. In response, therefore, a number of scientific advisors such as John von Neumann, George Kistiakowsky, and Jerome Wiesner in cooperation with Air Force representatives such as General Bernard Schriever and Assistant to the Secretary of the Air Force Trevor Gardner, began to push the Eisenhower Administration for an expanded missile program.[14] Thus would be ushered in a new phase in the arms race.

A second change in the arms race was that nuclear technology appeared to many scientists to be reaching a relatively stagnant point. Although improvements in nuclear warheads would continue to be possible through research and development, such possibilities were regarded by many scientific advisors as less important than other priorities such as arms control measures, continental defense, and, presumably, other projects advocated by the ultra-secret report in 1955 of the Technical Capabilities Panel (Killian Committee) of the

[13] Robert Oppenheimer, "Atomic Weapons and American Policy," *Bull. Atom. Sci.,* Vol. 9, No. 6, July 1953, p. 203.

[14] U.S. House of Representatives, Committee on Government Operations, *Organization and Management of Missile Programs,* Hrpt. 1121 86th Congress, 1st Session, 1959, p. 70.

Science Advisory Committee to the Office of Defense Mobilization.[15]

In the light of these shifts in the nature of the East-West arms race the testing of new designs of nuclear warheads warranted, in the eyes of many experts, a lower priority than it had in the past. A start toward some means to limit the nuclear arms race and to establish strategic stability was believed to be a far more important goal for American policy.

THE APPARENTLY DECREASING TECHNICAL FEASIBILITY
OF DISARMAMENT

At the same time that disarmament appeared to many to be more desirable than a continued arms race, the arrival of nuclear parity introduced a new complicating factor. If both East and West should produce great stockpiles of nuclear weapons, would disarmament ever be possible? Could East or West ever assure the other that it had no hidden stockpiles of weapons with which it could destroy the other?

Long before 1955 most scientists had been aware that the original disarmament proposal of the United States, namely the Baruch Plan, was no longer technically feasible; now they were realizing as well that the technical feasibility of *any* disarmament system would decrease more and more rapidly in the future. Without a moratorium on the continued development of novel delivery and defense systems some scientists foresaw that the world would reach the point where nations

---

[15] There is virtually no public information available on this panel which is otherwise known as the Killian Committee. Like the Gaither Committee (see Chapter Six), it reviewed the national defense posture and advocated the initiation or expedition of numerous projects. For example, the panel without doubt advocated a greater effort in continental defense and probably stressed the need for accelerating the missile program. Unlike the Gaither Committee the existence of the panel and of its report was kept secret until much later. This may account in large part for the reported success of the committee. In any event the excellent work of the panel's chairman, James Killian, must have become a significant factor in his later selection by President Eisenhower as his Special Assistant for Science and Technology.

would face each other with nuclear tipped missiles hidden throughout the landscape and where these nations would be wholly dependent for their security in limited engagements on tactical nuclear weapons. In that case it might become technically impossible ever to disarm atomically as there could be no guarantee that any inspection system would uncover hidden stockpiles of missiles and nuclear weapons. It would be possible for one party to a treaty to hide caches of atomic weapons at the same time that another disarmed in good faith.

The only real hope for total disarmament lay in the possibility that the further development of nuclear weaponry could be arrested in the near future. Therefore, in 1955, the world seemed to many scientists to be standing at the parting of the ways once again. In one direction lay a world slowly retreating from the mutual deterrence of a balance of terror to the security of permanent disarmament. In the other direction lay a world permanently armed with weapons of terrible swiftness and destruction. The time for decision was short and once past there might be no hope of return.

THE PROBLEM OF THE NTH POWER

In the Franck Report scientists had warned that as the knowledge of atomic energy was expanded, diffused, and utilized the number of nuclear powers would increase. If the number of these powers did increase it was believed the stability of the world would decrease correspondingly. Many scientists, greatly concerned over this possibility, discussed it first as "the problem of the spread of atomic energy," then as "the problem of the fourth power," and finally as "the problem of the Nth power." They have believed that the threat to the world multiplies rapidly as N increases.

In the ten years since the Hiroshima explosion, two nations, Great Britain and the Soviet Union, had joined the nuclear club. By 1960 a fourth power, France, was to join the decreasingly exclusive club, and other nations were standing on the threshold. Given the industrial and scientific potential of other nations, the number could be increased to at least a dozen by 1970.

During the intense period of the Cold War, from 1947 to 1955, the concern of scientists over the problem of the Nth power had been overshadowed for the majority by the blatant threat of Soviet aggression. Obviously the greatest need in the West then had been the development of a military power which would balance the Soviet power on the Eurasian continent. This had meant the development of tactical atomic weapons to overcome the Western deficiency in firepower, and since Project Vista the train of events had led to the point where a decision had been reached in December 1954 that NATO strategy would be based on tactical atomic weapons. By 1955 it was feared by many scientists that unless atomic weapons were brought under some system of international control, it would be just a matter of time before the individual European nations themselves and then non-European nations would possess an independent nuclear capability.

Now, in the changed world of the "Soviet Thaw," nuclear parity, and the decreasing technical feasibility of disarmament, the problem of the Nth power reasserted itself. The same scientists in the finite containment school who had formerly supported the sharing of nuclear weapons with America's European allies came to oppose the policy of creating an independent European nuclear deterrent. Although the short range result of that policy would be a strengthening of the hand of the West after arrival of nuclear parity, in the long run this policy seemed to lessen the chances for eventual disarmament and to decrease world stability. Furthermore, these scientists believed that the self-interest of the United States and the Soviet Union in preventing the rise of new nuclear powers would draw them together to work towards an acceptable control plan over nuclear weapons.

The Response of the Control and Finite Containment Schools: The "First Step" Philosophy

In the light of all these changes, since 1955 the scientists in the control and finite containment schools have felt that a renewed attempt must be made to solve the problem of atomic weapons, and the majority within the scientific community has supported this view. The conviction of the finite containment scientists, expressed in the Franck Report, that the real obstacle to international control of nuclear weapons is the lack of mutual trust has again become dominant in their thinking about international politics; thus they believe that their primary social responsibility today is to convert the nations to faith in control.

Their hope that such a conversion of the nations could take place was strengthened by the mounting evidence that the scientific community in the Soviet Union was at last obtaining its freedom. After 1955, Russian scientists were not only again appearing in the West but they were being accorded great prestige by Soviet society. In fact, many scientists were known to occupy positions of high influence in the Communist hierarchy.

This development rekindled in the hearts of a great many American scientists, particularly the physicists and other scientists who constituted the leadership of the finite containment and control schools, a very powerful faith that science as an international community and language is a great force for peace. Many of these scientists who were European born or trained had been reared in the pre-World War II atmosphere of international science, and the apparent possibility of reestablishing the bonds of international science through new contacts with Soviet scientists seemed to them to be a momentous opportunity. The greatest need, as the editorialist for the *Bulletin of the Atomic Scientists* had written as early as 1946 is " . . . broadening and internationalization of the fight for a rational solution of atomic energy problems. We need an international organization of scientists dedicated to this fight."[16]

At the same time that American scientists were awakening to the possibility of renewed contacts with Soviet scientists and were beginning to see in this a dim hope for peace, Administration thinking was also undergoing a modification. Upon the initiative of the Atomic Energy Commission it was decided that increased contacts between scientists of the East and West might serve to stimulate Russian interest in President Eisenhower's Atoms for Peace program. Specifically, the Administration proposed a Conference on the Peaceful Uses of Atomic Energy which was held in 1955.

The impact on the thinking of American scientists of this conference where for the first time there had been a "massive confrontation of East-West scientists" was expressed by Isador Rabi, a strong mover behind the conference. In 1960 Rabi— believing science to be a force for peace—wrote of the conference as follows: "It opened an era of good feeling, of mutual understanding and respect, between these two groups of scientists, which has increased in time as more Americans have had the opportunity of visiting Russian installations, and vice versa. The intervening five years have shown that science and peace do have a connection, not yet as great as science and war but certainly not negligible. . . . Indeed, one can say that science has provided an important bridge between the two rival blocs and a means for further peaceful cooperation, for the lessening of tension and suspicion."[17]

In time, these scientists believe, the workings of Soviet science will destroy the dogma which underlies Soviet hostility and suspicion. Consequently, the task for American scientists

[16] Editorial: "International Cooperation of Scientists," *Bull. Atom. Sci.,* Vol. 2, Nos. 5 and 6, September 1946, p. 1.

[17] I. I. Rabi, "The Cost of Secrecy," *The Atlantic,* Vol. 206, No. 2, August 1960, p. 39. For a history of the conference see Laura Fermi, *Atoms for the World,* The University of Chicago Press, 1957.

is to encourage science in Russia as a force for peaceful cooperation. Every opportunity has to be found to increase contacts and to open up the Soviet Union, particularly through establishing the fact that Western and Soviet scientists can meet together and lay a technical foundation for the resolution of the problem of nuclear weapons. Scientists can and should lead the world out of the perils of the atomic age just as they had led it in.

Underlying the view of these scientists is the strong conviction that, in the face of probable destruction, a nation which is sufficiently advanced in science to produce hydrogen bombs is rational enough to realize the overriding imperative of disarmament, even though this requires that it forsake major national goals. This conviction is reinforced by the equally strong belief that science is acting within each nation as a rationalizing force.

In essence, what these scientists wanted after 1955, then, was a new approach to the problem of disarmament like that made in 1945. The United States ought to elaborate a disarmament plan consistent with the contemporary technical facts and to present it to the Soviet Union. They believed that in time the merits of such a plan would commend it to the Russians. Rather than desiring to wait for the Soviet Union to indicate a willingness to bring atomic energy under control before initiating novel plans, these scientists hoped that the creation of a feasible plan in itself would instill in the Russians a willingness to negotiate the problem. David Inglis, who was to become President of the Federation of American Scientists, wrote as early as 1952 in this vein: *"We can't be sure that they [the Soviets] will never agree to any and all significant inspection plans until we have presented all plans that could ameliorate the situation, and propounded them with emphasis on the incentives."*[18]

[18] David Inglis, "Tactical Atomic Weapons and the Problem of Ultimate Control," *Bull. Atom. Sci.,* Vol. 8, No. 3, March 1952, p. 81. (Italics his.)

Specifically, these scientists in 1955 desired that the United States propose to the Soviet Union a "first step" to the solution of the problem of nuclear weapons. The "first step" which they now suggested was a nuclear test ban, as it was held to be technically feasible and was not expected to involve serious political obstacles; scientists expected that the success of such a step would rebuild confidence and mutual trust between the Great Powers. Agreement and confidence then would follow one another until the other steps to total disarmament had been taken. For all these reasons the idea of a nuclear test ban gave a focus for action to people who wanted to do something about the nuclear arms race.

Whereas a ban on the testing of megaton weapons had been under discussion in the Administration since the spring of 1954 and would become a major issue in the 1956 presidential campaign, a nuclear test ban as a "first step" disarmament agreement was a novel idea. The belief developed that a ban on *all* nuclear weapons tests (like that proposed by Prime Minister Jawaharlal Nehru of India in 1954) would not only be the first concrete accomplishment in a decade of disarmament negotiations but it would serve to symbolize to a discouraged world that the cause of disarmament was not lost. Furthermore, it would encourage the Great Powers to take other steps toward eventual complete disarmament.

This philosophy of the "first step" found its clearest and most forceful expression in a petition drafted by Linus Pauling and signed by over 9000 scientists which was delivered to the United Nations in January 1958. "An international agreement to stop all testing of nuclear weapons now," the petition read, "could serve as a *first step* towards a more general disarmament, and the effective abolition of nuclear weapons, averting the possibility of a nuclear war that would be a catastrophe to all humanity."[19]

[19] Quoted by Noel-Baker, *op.cit.*, p. 263. (Italics mine.) For a discussion of U.S. policy on a nuclear test ban prior to 1958 see Thomas Murray, *Nuclear Policy for War and Peace,* World Publishing Co., 1960, Chapters 4 and 5.

Under this banner of a "first step" disarmament agreement, the control school re-emerged and the finite containment school repudiated its previously supported program to further the development of tactical nuclear weapons. These two schools joined in an advocacy of a series of technical solutions to various aspects of the nuclear weapons problem; they believed the first step in this direction might be a total ban on all nuclear weapons tests. In this advocacy they were to be strongly opposed by the infinite containment scientists.

THE RE-EMERGENCE OF THE CONTROL SCHOOL

In the period from 1947 to 1954 the control school of thought had been negligible in its influence. On occasion scientists like Philip Morrison, Harlow Shapley, and Linus Pauling had issued statements and the pacifist Society for Social Responsibility in Science had been formed during this period. On the whole, however, the ideas of these men had carried little weight. The intransigence of Stalin, the obvious need for the United States to diversify its weapons program, and the influence of Robert Oppenheimer had held them in check. Their view that with a little flexibility and understanding from the United States the Soviet Union would soften its policies had been obviously contrary to the stark realities of the Cold War. However, the changes which were taking place by 1955 made their position respectable and tenable once again. Of these changes, the discovery of the genetic and pathological effects of radioactive fallout from weapons testing provided the greatest impetus for the revival of this school.

The control school recruited scientists who believed that, given the high probability of these health dangers, it was insane and criminal to continue bomb tests. As an outstanding biologist told this writer, "those irresponsible physicists are insane with their building of bigger and bigger bombs." This belief is characteristic of the great number of scientists who came to accept the view of the control school. It holds that

American bombs are big enough already and there is no need to continue testing. In fact, they view it as dangerous to test or to build bigger bombs because of their pathological and genetic effects.

The new recruits to the control school were primarily from the biological sciences, and with a few exceptions, had not previously been involved in the struggle for the international control of atomic weapons. They were scientists who were acquainted with neither the specific and multiple purposes of testing nor the complex problems associated with control. As a result, their demands for control have tended to be extreme because they do not know of the great difficulties involved of which the finite containment scientists are aware.

The control school commenced its renaissance in London during July of 1955, the same month in which the Geneva Summit Conference was held. In an appeal to the world, which resulted in the Parliamentary Association for World Government, Bertrand Russell and Albert Einstein pointed out the dangers of thermonuclear war. This appeal was supported by another signed by fifty-two Nobel Laureates asking that the nations renounce force as a final resort of policy. The third and high point in these appeals came in January 1958 with an attempted *tour de force* when Linus Pauling presented a petition to the United Nations which urged the immediate cessation of all nuclear tests and had been signed by 9235 scientists from many nations.

The 1955 meeting of the Parliamentary Association for World Government gave rise to a number of other movements supporting the international control of atomic energy and an immediate suspension of nuclear tests such as the Pugwash Movement and National Committee for a SANE Nuclear Policy.[20] Of these, the more noteworthy is the Pugwash Move-

[20] An interesting treatment of SANE is Nathan Glazer's "The Peace Movement in America," *Commentary*, Vol. 31, No. 4, April 1961, pp. 288–96.

ment,[21] named after its first meeting grounds, the summer home
of sponsor Cyrus Eaton, a Cleveland industrialist. The significance of this movement lies in its international character, as it
is the first such movement in which Soviet scientists have
actively participated. The Pugwash Movement is essentially
an attempt to utilize the internationalism of science as a force
for peace. In a world divided by nationalistic and ideological
loyalties, the Movement's sponsors hope that science as an
international community and language can provide a means
of communication and peaceful accommodation between
East and West.

Lastly, the position of the control school is based on the
belief that the arrival of mutual deterrence and the dangers
of radioactive fallout have made nuclear war, if not all war,
impossible for reasonable and informed people. The control
school sees nuclear war as an impossibility providing that
reason prevails, because both the United States and the
Soviet Union are forever deterred from waging war. "I believe," wrote Linus Pauling in the beginning of his book,
No More War!, "that there will never again be a great world
war, if only the people of the United States and of the rest
of the world can be informed in time about the present world
situation. . . . I believe that the development of these terrible

[21] The character of the Pugwash Movement, whose 1961 meeting at
Stowe, Vermont, was known as the International Conference on Science
and World Affairs, has changed in its convocations over the past six
years. It has now spurned the support of Eaton and others whose past
associations have prejudiced opinion against the movement; its scientific
leadership has shifted from the control to the finite containment school;
it has become more conscious of and has succeeded in avoiding exploitation by Communist propaganda; it has become more modest in its ambitions as a force for peace.
Unfortunately there is no thorough history of the Pugwash Movement. There are, however, numerous articles in the *Bull. Atom. Sci.* In
addition there is a very critical study by the Internal Security Subcommittee of the U.S. Senate, Committee on the Judiciary: *The Pugwash
Conferences,* A Staff Analysis Prepared for the Subcommittee to Investigate the Administration of the Internal Security Act and other Internal
Security Laws, 87th Congress, 1st Session, 1961.

weapons forces us to move into a new period of peace and
reason, when world problems are not solved by war or by
force, but are solved by the application of man's power of
reason, in a way that does justice to all nations and that
benefits all people."[22]

The theme that the nuclear arms race is due to misunderstanding, ignorance, and irrationalism pervades the approach
of these scientists to the problems presented by nuclear weapons. It is believed that there is a great conflict between the
vestigial forces of irrationalism and the progressive force of
reason within each of the major powers. The immediate task
set for the scientists and others is the encouragement of the
forces of reason within the United States; once the United
States has been won over to the cause of reason, it will then
be able to encourage the progressive groups in Soviet society.
Eventually East and West, following the dictates of reason,
will be able to find a negotiated settlement of the nuclear
arms race.

THE NEW POSITION OF THE FINITE CONTAINMENT SCHOOL

Although the scientists in this school tend to subscribe to
the thesis of the control school presented above, they are far
more cognizant of the problems involved in ending the nuclear
arms race. Their faith that if man's reason would only assert
itself the arms race could be solved is tempered by a keen
awareness of the inherent difficulties of disarmament and the
issues dividing East and West. There is a clash in their
thinking between their desire to end the arms race and their
realization of the immensity of this goal.

The membership of the finite containment school continues
to be made up largely of scientists who were members of the
Manhattan District Project or those younger scientists who
were influenced by the former Project scientists in earlier
attempts to secure international control of atomic energy.
Their contacts with the Atomic Energy Commission, their

---

[22] Linus Pauling, *No More War!*, Dodd, Mead, and Co., 1958, p. 3.

participation in government, and their experiences in attempting to achieve international control have made them
at least relatively appreciative of the political problems involved in disarmament. However, the events just prior to and
since 1955 have given them the impression that the new
conditions in the world have made a "first step" toward
disarmament both feasible and desirable.

Hans Bethe, a leader in this school, has frequently stated
the faith that one day the Soviets will see that international
control is to their interest. The Soviet thaw indicated to him
that the Russians might already be more favorably disposed
to an agreement. If so, the United States should take advantage of the new opportunity to halt the arms race. "I believe,"
Bethe told the Senate Subcommittee on Disarmament, "that
the most important point [in achieving a test suspension] is a
political one; namely, to obtain a controlled disarmament
agreement, and I think this is an objective which the United
States has had ever since the war, and it never has come so
close to realization as in this area. And if we once get one
controlled disarmament agreement, I believe that others may
follow, and that the principle will thereby be established. This,
I think, is the overriding argument."[23]

The finite containment scientists believe that, in addition
to its immediate consequences, a test ban would serve three
very significant purposes. Firstly, it would symbolize to the
world the determination of the United States to end the arms
race. Secondly, it would serve to improve the climate of
international relations. The " . . . inspection system . . .
would be of value, because it would to some extent open up
Russia and help to reduce tensions between the two countries."[24] Thirdly, a test ban would preserve mutual deterrence,
prevent the rise of new nuclear powers, and thus stabilize the
world.

[23] U.S. Senate, Committee on Foreign Relations, *Hearings before a
Subcommittee on Disarmament and Foreign Policy*, 86th Congress, 1st
Session, 1959, p. 178.
[24] *ibid., Hearings . . . on Control and Reduction of Armaments*, 85th
Congress, 2nd Session, 1958, p. 1544.

For all these reasons the finite containment school reversed its former position on the further development of tactical nuclear weapons and gave its support to a nuclear test ban. The changes in the world made the price to be paid for increasing the flexibility of America's nuclear capabilities appear too high to these men. This shift brought the finite containment group into tacit alliance with the control school and into open conflict with the infinite containment scientists who remained committed to the desirability of maximizing America's nuclear capabilities.

Scientists Opposed to the "First Step"; The Infinite Containment School

The infinite containment school position is that even if a test ban were feasible the possible gains from a ban would be negligible compared to the sacrifice which it would require of the United States. A test ban would prohibit the United States from developing the tactical atomic weapons which these scientists believe it needs to meet the overwhelming Soviet advantage in conventional power and it would increase pressures to outlaw all atomic weapons as desired by the Soviet Union.

At the same time, they argue, a ban would not necessarily prevent the Soviet Union from secretly developing improved atomic weapons. This is due to the fact that any control system would be developed on the basis of an unstable body of knowledge, and as knowledge expanded novel methods of infringement would evolve. In fact, contrary to the view of the proponents of a test ban that a ban would be advantageous to the United States because the United States leads the Soviet Union in atomic weapons, these scientists believe that a test ban would give the Soviet Union an opportunity to overtake the United States. The decisiveness of modern weapons, the importance of lead-time, and the delicacy of the balance of power place "a competitive premium on infringement." Furthermore these men assert that if the Soviet

Union should achieve a year's lead over the United States in new types of atomic weapons, mutual deterrence might vanish and the balance of power could shift decidedly in the Russian favor. The possibility of achieving this military advantage would be too great an incentive for the Soviet Union to be deterred from attempts to achieve it by a fear of being caught. This very dangerous risk that a test ban may create an opportunity for the USSR to shift the balance of power in its favor is not believed to be compensated for by the alleged advantages of a nuclear test ban. Consequently the infinite containment scientists maintain that until the Soviet Union becomes an open society, partial systems of international control over atomic energy will be ineffective and in fact dangerously misleading.

Conclusion

This chapter provides the background for the intense conflict over nuclear weapons testing that was to begin in 1956 and to continue to the present day. While to the public the debate would appear to be primarily concerned with the harmful effects of radioactive fallout, the real issue dividing the scientists—as became clear after 1958—was that of the wisdom of ending the nuclear arms race. This is not to say that the question of fallout has never been important; rather it is to say that this was not the burning issue that really divided the scientists. That issue was, at least for the leaders in the debate, whether or not it was wise to pursue the philosophy of the "first step."

CHAPTER SIX

THE VICTORY OF THE "FIRST STEP" PHILOSOPHY: THE GENEVA SYSTEM AND ITS CONSEQUENCES

DESPITE the growing pressures for a nuclear test ban as a first step toward disarmament, the Eisenhower Administration pursued a nuclear policy which assumed the continuation of the nuclear arms race and of nuclear weapons. In effect the doctrine of massive retaliation remained the cornerstone of American military policy while, at the same time, the Administration was creating through its testing program a capability to defend the West by means of tactical nuclear weapons. American disarmament policy reflected this dependence upon nuclear weapons and would continue to do so until after the advent of Sputnik in October 1957.

The Struggle over American Disarmament Policy (1955—Sputnik)

THE NEW AMERICAN POSITION: SUBSTITUTION OF ARMS CONTROL FOR DISARMAMENT

On March 19, 1955, just prior to the new Soviet Plan of May 10th, President Eisenhower had announced the appointment of Harold Stassen as his Special Assistant on Disarmament. Stassen's task was to fashion a new American disarmament plan consistent with changes in the world scene and with America's foreign policy goals. To assist him in this responsibility, Stassen assembled task forces to study various problems for which disarmament techniques would have to be worked out.

As head of the task force on inspection and control of nuclear materials, Stassen appointed one of the leading members of the infinite containment school, E. O. Lawrence,

Director of the University of California Radiation Laboratories. Lawrence had been one of the foremost advocates of the hydrogen bomb and shared the skepticism of his fellow infinite containment scientists concerning the feasibility of disarmament.

On the basis of the reports from Lawrence's and other task forces and with the concurrence of Administration leaders, Stassen formulated a new American position on disarmament. The first manifestation of this new American approach came at the Geneva Summit Conference (July 18–21, 1955). There, in his dramatic "Open Skies" proposal, President Eisenhower approached the problem of nuclear weapons indirectly. The Administration now believed that rather than seeking to implement an ambitious plan such as the Baruch Plan or the Russian proposal of May 10th, the world would have to take what the President referred to as "initial . . . confidence building measures" which would eventually lead to effective control over nuclear weapons. It believed specifically that the best way to build confidence would be through the establishment of measures by which to prevent surprise attack.

Commencing on August 29, 1955, Stassen himself began to reveal the new American position on nuclear disarmament. While the United States did not officially withdraw or disavow any of its former policies at this time, it did place a "reservation upon all of its pre-Geneva substantive positions."[1] And in effect this truly meant a reversal of American disarmament policy.

Prior to this "reservation," American disarmament policy, beginning with the Baruch Plan, had set as its goal the complete elimination of nuclear weapons from the arsenals of the nations. Now, on the basis of the failure of Lawrence's task force "to discover any scientific or other inspection method which would make certain of the elimination of nuclear

[1] U.S. Department of State, *Disarmament—The Intensified Effort 1955–58*, U.S.G.P.O., 1958, p. 18.

weapons," the United States set a new goal for itself in disarmament negotiations.

In place of nuclear disarmament, i.e., the total elimination of nuclear weapons, the United States now proposed that the nations should seek to preserve peace through mutual deterrence maintained by arms control measures such as the Open Skies proposal and the sharing of military blueprints. Rather than seeking to eliminate nuclear arms, the nations should try to control them.

This new position not only viewed mutual deterrence as a far more plausible course of action than the attempt to eliminate nuclear arms, but the latter was now considered to be a dangerous course to follow. "It is our view," Stassen told the London Disarmament Conference, " . . . that if armaments . . . are brought down to too low a level, then . . . instead of the prospects of peace being improved, the danger of war is increased."[2]

It should be noted in this connection that the Russians in their new plan proposed on May 10, 1955, had concurred with the "scientific facts" brought out by Lawrence's task force. Even though the delegates of the USSR did not draw the same conclusions as did the United States from these scientific facts, the Russians had made an admission in their new plan of profound significance for future disarmament talks. This was their recognition of the problem of accounting for nuclear stockpiles. The Russians had stated: " . . . There are possibilities beyond the reach of international control for evading control and for organizing the clandestine manufacture of atomic and hydrogen weapons, even if there is a formal agreement on international control. In such a situation, the security of states signatories to the international convention cannot be guaranteed, since the possibility would be open to a potential aggressor to accumulate stocks of atomic and hydrogen weapons for a surprise atomic attack on peace-loving

[2] Quoted in Philip Noel-Baker, *The Arms Race,* London: Stevens and Sons, 1958, p. 29.

states."[3] Despite this admission, however, the USSR continued to press for the total elimination of nuclear weapons.

THE CONFLICT OVER THE NEW AMERICAN POSITION
ON DISARMAMENT POLICY

This new position of the Eisenhower Administration on disarmament policy struck at the heart of the growing conviction held by many scientists in the finite containment and control schools that nuclear weapons *had* to be eliminated. While they accepted the fact that science then knew no method to uncover stockpiles of nuclear weapons, these scientists did not believe this would necessarily always be true. Moreover, the problem of undetectable stockpiles did not, in their view, warrant the conclusion that the elimination of nuclear weapons was a hopeless cause. Instead the search for a solution to the problem ought to continue.

All arms control measures seemed to them to be temporary expedients which could do no more than stabilize the balance of terror and thus give the nations time in which to seek elimination of nuclear weapons. Moreover, the suggestion that permanent peace be based on a "balance of terror" was anathema to these scientists. In the long-run, they argued, it would not work.

The first major skirmish between these scientists and the supporters of the new American position came in the presidential election of 1956. Although the "first step" philosophy had yet to be articulated clearly, the idea that a suspension of tests would be a good thing was already current among many scientists such as those in the leadership of the Federation of American Scientists. At the least, they thought, the testing of thermonuclear weapons should cease as such tests not only caused harmful fallout, but it also seemed evident that the perfection of these weapons would make possible the development of an intercontinental ballistic missile (ICBM) to the detri-

[3] U.S. Department of State, *Disarmament—The Intensified Effort*, *op.cit.*, p. 15.

ment of the United States. Many of these men expected that the ICBM with a hydrogen warhead would benefit the USSR far more than the United States which could already strike the Russians from its overseas bases. Also, the worldwide concern over fallout was harming the prestige of the United States among the uncommitted nations. Lastly, such a ban would slow the nuclear arms race somewhat.

These scientists found a champion in Adlai Stevenson, the Democratic candidate for President, and the wisdom of the continuation of hydrogen bomb tests became a major issue in the presidential campaign. Although raising an issue as complex as this one undoubtedly helped to defeat Stevenson, it did fire interest in the subject among more scientists as well as others.

Following the defeat of Stevenson, the issue of nuclear testing lay dormant for a few months. Commencing in early 1957, however, this relative calm was shattered by numerous events, all of which would increase the pressure on the United States for the cessation of *all* nuclear tests. On April 30, 1957 the USSR proposed that the question of nuclear tests be isolated from other disarmament issues for separate treatment by the nuclear powers and that it be settled "without delay."[4] Also, the humanitarian plea of Albert Schweitzer that testing cease because of its harmful effects had a considerable impact on the American conscience. Of profound significance as well was Linus Pauling's petition signed by thousands of scientists calling for a nuclear test ban as a first step to general disarmament. And Pauling, who carried his fight to the United Nations and throughout the United States, tried to insure that he be heard and that his petition be acted upon.

Throughout this public controversy the leaders of the infinite containment school such as Edward Teller and Willard Libby acted as the Eisenhower Administration's first line of defense for its testing policy. Supported by Lewis Strauss, these

[4] *ibid.*, p. 38.

scientists defended the increasingly unpopular policies of the Administration for which they themselves were largely responsible. The specific issue upon which the battle was joined was that of the danger to humans of radioactive fallout.

The rancor and bitterness of the conflict among the scientists over the dangers of radioactive fallout was unparalleled in the whole history of scientific conflict over nuclear weapons policy. While the chief protagonists in the struggle, Linus Pauling and Edward Teller, remained constantly within the bounds of the limited scientific evidence,[5] their interpretations of this evidence ranged on occasion from the one extreme that fallout was equivalent to the plague to the other that fallout might even be beneficial.

An example of the way in which the debate was handled by the chief protagonists is found in the use by each side of the same data (from the research of Dr. Hardin Jones of the Donner Laboratory, University of California at Berkeley) on the harmful effects of fallout. Whereas Pauling presented the danger in absolute terms, i.e., the number of individuals (especially children) who would probably die due to fallout, Teller presented it in relative terms, i.e., the number of days lost per life for the American people as a whole due to fallout relative to the shortening of life due to smoking.[6]

Nevertheless the extremes to which the debates went and the apparent manipulation of scientific data for partisan purposes were such as to upset much of the scientific community. Equally disturbing was the view of many scientists that each side was guilty of intellectual dishonesty. In particular, many physicists believed that Pauling's whole position was dishonest,

[5] For a discussion of this point see Eugene Rabinowitch, "Decision Making in the Scientific Age" in Gerald Elbers and Paul Duncan, eds., *The Scientific Revolution,* The Public Affairs Press, 1959, pp. 24–25.

[6] Interview with Dr. Hardin Jones, May 24, 1959. For a critical and penetrating discussion of the debate over fallout, see James McCamy, *Science and Public Administration,* University of Alabama Press, 1960, pp. 185–96.

that Pauling's campaign against testing was really based on pacifist motives, and that the issue of fallout was a subterfuge seized by him as a device to alarm public opinion. Pauling's critics charged him with making alarmist predictions about the dangers of fallout in order to create a public opinion against *all* nuclear testing.

Many of these same critics also turned their wrath against the leaders of the infinite containment school because they were believed to have joined with Lewis Strauss to deceive President Eisenhower and the American people on the matter of "clean" nuclear weapons. On June 24, 1957, at the suggestion of the Joint Committee on Atomic Energy, Teller, along with Ernest Lawrence and Mark Mills, told President Eisenhower that the United States knew how to make 95 per cent clean hydrogen bombs, i.e., bombs in which 95 per cent of the radioactivity is eliminated, and that with further testing it might be possible to make "smaller hydrogen bombs with essentially no radioactive fallout."[7] They implied—or so it seemed to their critics —that it might be possible to reduce the size of these "clean" hydrogen weapons to that of "nominal" atomic bombs and that such weapons would be suitable for employment in tactical situations.

The critics of Teller and his colleagues angrily made the accusation that while these statements were technically correct, they led to an erroneous conclusion and were part of a concerted effort to prevent a nuclear test ban agreement. The statements implied, these critics argued, that the United States was on the verge of a major breakthrough in weapons technology which, with further testing, would enable the United States to produce "clean" small-yield tactical atomic weapons. These critics argued instead that 95 per cent "clean" applied only to large-yield weapons where the fission trigger is small compared to the total yield of the weapon or to weapons det-

[7] *The New York Times,* June 25, 1957, p. 1.

onated at high altitudes.[8] Barring the development of a pure fusion weapon, which the most vociferous critics of the President's advisors argued was "contrary to the laws of physics," they denied the possibility of a fallout-free or "clean" bomb.[9]

And, even if pure fusion tactical weapons were developed, these critics doubted that such weapons would be as significant militarily as the proponents of continued testing claimed. This was because the use of such weapons in tactical situations would require detonation close to the ground in order to limit the area destroyed. However, if the weapons were exploded close to the ground, the explosion would scoop up great quantities of earth and create vast radioactive clouds of debris. Yet if exploded at a high altitude, the critics argued, such "tactical" weapons would become indistinguishable from strategic weapons.

Whatever the merits of the various arguments, Pauling's campaign against testing on the grounds of its harmful results was to have a profound effect on American nuclear policy. Public concern that testing was releasing into the atmosphere deadly agents of leukemia and causing genetic damage placed

[8] A technical note is perhaps necessary here. Under present technology, thermonuclear (or fusion, or hydrogen) weapons have a fission trigger and outside casing. It is the fission trigger that accounts for most of the radioactive debris from such weapons. The absolute quantity of this fissionable material in any thermonuclear weapon tends to be a constant; as the yield of the weapon increases, therefore, the relative "cleanness" of the weapon improves.

A third source of radioactive fallout is the contact with the earth of the explosion's fireball as the scooped-up earth becomes radioactive. Presumably, if a pure fusion weapon were developed and detonated so that its fireball did not touch the earth, there would be little radioactive debris.

[9] Secrecy prevented an adequate public discussion of this new concept in atomic armament. The issue would be raised again in the spring of 1960 by Senator Thomas Dodd and physicist Freeman Dyson. Together they amplified the scanty details on what has become known as the "neutron bomb." Only in 1961 would the neutron bomb become well-known to the public at large; yet the actual nature of the bomb and of what the argument was all about would remain unclear. See, for example, *The New York Times*, November 2, 1961, p. 2.

a heavy moral burden on the Eisenhower Administration. This and other pressures for a total ban on nuclear weapons as a first step toward nuclear disarmament were having a noticeable effect. In addition, Harold Stassen, the President's Disarmament Advisor, was reaching the conclusion that the path to some significant agreement on nuclear weapons with the Russians could be taken only following the first step of a nuclear test ban. Along with others in the government he began to support this course of action.

In order to meet the mounting criticism of its policies and to shift the responsibility for continued testing to the Russians, the United States at the London Disarmament Conference proposed a new "package" arms control plan on August 29, 1957. In this plan the United States accepted the Russian position that a test ban should be the first disarmament proposal to be negotiated and agreed that it would accept a two year suspension of nuclear weapons tests provided the Russians agreed to and permitted, during the period of test suspension, the installation of an international control system to "cut off" all future production of nuclear materials for weapons purposes.[10]

Although this package plan contained a number of interesting proposals such as control over outerspace missiles, two points merit special attention. In the first place, the United States maintained the position first taken in late spring 1955 that the maintenance of mutual deterrence and not the elimination of nuclear weapons was the goal of American disarmament policy. Secondly, the United States in effect argued that it would not cease its nuclear tests unless the USSR agreed to an opening of the Iron Curtain; specifically any inspection system to insure only peaceful uses of future nuclear production would have to be as stringent as that proposed in the Baruch Plan.

At this time the United States still held to the view that a

[10] U.S. Department of State, *Disarmament—The Intensified Effort, op.cit.,* pp. 48–50.

test ban would interfere with American weapons plans, and this package was tantamount to saying that the United States could not agree to a test ban unless the Soviet Union became an open society. The violent Soviet reaction to this suggestion brought to an end this phase of the disarmament negotiations and on September 6, 1957, the USSR withdrew from what it termed "the fruitless disarmament talks."[11]

In summary, as of the fall of 1957, the Eisenhower Administration was strongly committed to the continued development and testing of nuclear weapons. Until the USSR made major concessions the United States would not cease testing. Then a Soviet technological advance upset American policy deliberations for the third time in a decade. On October 4, 1957, the news reached the United States that the Soviet Union had launched the first artificial earth satellite. This was an event of momentous importance. And although the President would insist as late as December 15, 1957, in a letter to Prime Minister Nehru that a test ban could not be "an isolated step," Sputnik did set in motion a new train of events.

The Impact of Sputnik on the Eisenhower Administration

Despite the immediate chorus of statements by Administration officials designed to calm the American people, there is little doubt that President Eisenhower and his advisors considered Sputnik a significant new power factor which indicated that decisive steps must be taken to redress the balance of power now likely to shift in the Soviet favor. Nonetheless, in retrospect it now also appears that the steps taken to redress the shifting balance of power were in certain respects inadequate to meet the Soviet challenge.

The lack of decisive action by President Eisenhower is all the more regrettable because, by coincidence, in the months preceding Sputnik a thorough evaluation of American defense posture had been in preparation for him. This was a study by

[11] *ibid.*, p. 51.

a group of private citizens many of whom were scientists and all of whom were experts in military matters. The report of these experts, which was given the name of the group's chairman, Rowan Gaither, was presented to President Eisenhower exactly one month after Sputnik.[12]

The fundamental purpose of the Gaither Report was to stress the danger to the United States of a Soviet surprise attack and to advocate measures to make America's retaliatory power invulnerable to attack. A second important point in the report emphasized the need for the United States to rebuild its conventional limited war capability. The United States, according to the report, was unprepared to engage in conventional limited war despite the fact that the strategic nuclear balance would make limited wars more likely, particularly in the Middle East and Asia. A third suggestion which is of interest here was that President Eisenhower should communicate to the American people the seriousness of the threat facing the nation and the need to increase defense expenditures. The members of the Gaither Committee were sure that, once informed of the dangers confronting the security of the United States, the American people would support a reinvigoration of the American defense effort.

For weeks thereafter, the President, impressed by the cogency of the arguments presented by such a prominent group of businessmen, scientists, and foreign policy experts, wavered on the verge of revising American strategic policy in a drastic fashion. President Eisenhower's "chins-up" speeches delivered in the winter of 1957–58 reflect his vacillation. In the end, however, following pressures from the Secretaries of the Treasury and of State, the President decided against major innovations in American military policy. The President did not believe the American people would support the needed acceleration in defense spending.

[12] This discussion is based primarily on a study by Morton Halperin which is a thorough piecing together of the scattered available information on the report. "The Gaither Committee and the Policy Process," *World Politics*, Vol. 13, No. 3, April 1961, pp. 360–84.

While certain steps toward defense reorganization and acceleration of the missile program were taken in response to the Gaither recommendations, the United States did little to increase its civil defense program or to develop an adequate capability for conventional limited warfare. Instead massive retaliation remained the essence of American military strategy despite the increasing Soviet strategic capabilities and its traditionally strong conventional power.

The principal American move following Sputnik which is of concern to this study was the increase in the integration of tactical atomic weapons into NATO forces. In his State of the Union message on January 9, 1958, President Eisenhower proposed that the Atomic Energy Act of 1954 be amended to authorize agreements on the exchange of atomic weapons information and materials with America's European allies. After a long period of debate, Congress passed, and on July 2, 1958 the President signed, the law permitting "agreements for cooperation for mutual defense purposes." Although in principle this act signified a major move toward nuclear sharing with America's European allies its practical effect would be much less. While it appeared that the United States could now share non-nuclear parts of atomic weapons, special nuclear materials for use in atomic weapons, and secret information on atomic weapons with its allies, in actuality the sharing of really significant information and materials was restricted by congressional mandate. In order to prevent European nations from obtaining independent nuclear capabilities Congress permitted the sharing of the most important data and materials only with nations which had made "substantial progress of their own." namely Great Britain. However, America did offer its European allies intermediate range ballistic missiles (IRBM), a move which actually ran counter to the Gaither Committee's admonition against dependence upon a vulnerable deterrent such as the type of IRBM offered to Europe.

Another result of Sputnik was an acceleration of the development of tactical nuclear weapons. The AEC scheduled a series of nuclear tests for the late summer and fall of 1958. Included in this series were to be some underground shots. The idea of underground tests had been proposed by Edward Teller and David Griggs in 1956 for scientific reasons. The first underground shot had been the Rainier shot of 1.7 kilotons in the fall of 1957. The fall 1958 series was code-named Hardtack.

As announced by the AEC, the purpose of the new test series was to demonstrate the feasibility of developing small size, low yield atomic weapons, some of which could be carried into battle by infantry.[13] In this connection the most important test apparently would be of the nuclear warhead for the Davy Crockett bazooka, an anti-tank weapon. While the power of the Davy Crockett remains classified, a statement by the late Thomas Murray may be pertinent. Murray, who had been a member of the AEC, wrote: "The development of a quarter-kiloton bomb, which might conceivably be fired from something like a bazooka, would place in the hands of a small team of ground fighters blast charges each on the order of 250 tons of TNT, or equivalent to the explosive power carried by some fifty B-29's in World War II."[14]

While these tests were being planned the changes in the international scene which had already so profoundly affected the scientific community and converted the majority of scientists to the support of a nuclear test ban as a first step to disarmament were being increasingly felt in various ways at the White House level. The tide of world opinion on the matter of fallout was running against the United States; the Soviet proposal to ban all nuclear tests appeared to most nations to be a reasonable one when both East and West already had a

[13] There were at least eight explosions with yields of less than 100 tons equivalent of TNT. The smallest nuclear explosion was equivalent to .15 ton of TNT. Other yields were 100 tons, 92 tons, 84 tons, 83 tons, 57 tons, 36 tons, and 6 tons. All of these explosions are considerably smaller than the 20,000 tons yield of the Hiroshima explosion and the 15,000,000 tons Bravo shot. See *The New York Times,* March 11, 1959, p. 21, and July 31, 1960, p. 31.

[14] Thomas Murray, *Nuclear Policy for War and Peace,* World Publishing Company, 1960, p. 65.

capability to over-kill one another. Many Administration officials feared that one day such sentiments might lead to a serious propaganda defeat for the United States. Lastly, there was a growing concern within the Administration over the dynamics of the nuclear arms race.

This concern over the nuclear arms race was of many kinds. First of all there was the realization that an indiscriminate spread of nuclear weapons would pose a threat to the maintenance of bipolar deterrence upon which the United States was beginning to rest its hope for peace; some means had to be found to limit the number of new nuclear powers. This concern was no doubt becoming a matter of overriding importance, especially with respect to the possibility that Communist China might soon obtain a nuclear capability. In addition, however, and of indeterminant importance, was a growing anxiety over the cost of the nuclear arms race. Rather than level off as the New Look had anticipated, the demands of weaponry upon the economy were ever on the increase, as missiles replaced aircraft and then were themselves replaced by later-generation missiles.

Furthermore, as the implementation of Project Vista began to make nuclear weapons "conventional" the same economizing attitude apparently began to be applied to them as that which had previously been applied to conventional high explosive weapons and had led to the New Look. The concern over the cost of tactical nuclear weapons became acute in the Administration after 1955 as Army modernization heading toward highly mobile nuclearized forces for limited war began to receive serious consideration by the Joint Chiefs of Staff.[15]

---

[15] This discussion is based on Maxwell Taylor, *The Uncertain Trumpet,* Harpers, 1960. Taylor writes that massive retaliation began to give way in 1955 following the Soviet H-bomb (see p. 24). However, the costs of missiles after Sputnik became so high that limited war weapons began to suffer and the breakaway from massive retaliation was retarded (pp. 65ff.). In fact, Army leaders were to interpret President Eisenhower's decision to seek a nuclear test ban as a victory for massive retaliation over their doctrine of graduated deterrence, *The New York Times,* August 23, 1958, p. 2.

If, on the other hand, the arms race could be stabilized at this point through a nuclear test ban, the United States could save the high cost of weapons' modernization. A few years later, President Eisenhower told his press conference of this conviction: "I am of the belief that, if you could have now a ban on all [nuclear] testing that everybody could have confidence in, it would be a very, very fine thing to stop this, for this very reason, if no other: it is a very expensive business, to begin with. The very first bomb we produced I think cost America $2 billion or more before we ever, ever had the very first one. And since that time I don't believe that our—although you'd have to look this up—but I think our appropriations have never been below $2 billion a year. So it is an expensive business."[16]

A further factor which was to make all these pressures for a nuclear test ban come to fruition within the Administration had been introduced shortly after Sputnik with the appointment on November 7, 1957, of James Killian, President of the Massachusetts Institute of Technology, as Presidential Special Assistant for Science and Technology. This appointment brought to the highest level of government a person sympathetic to the finite containment position on nuclear weapons. Consequently, for the first time in a number of years, these scientists had an effective voice at the highest level of government.

In addition, the elevation of the Science Advisory Committee to the White House level brought in as presidential advisors a number of scientists of the finite containment school. For years these scientists had bided their time in the lower echelons of the bureaucracy while their political opponents in the infinite containment school exerted a strong influence on policy-making. During this time many of them had felt hostility toward Teller because of his part in the Oppenheimer security trial. Now, after the passage of just a few months, it would be

[16] Press Conference, *The New York Times,* February 4, 1960, p. 12.

these men who would advise the President and his chief aides on nuclear weapons policy.

In Killian and in the scientist-members of the President's Science Advisory Committee (PSAC), President Eisenhower found a group of brilliant and thoughtful men to whom he could turn for alternative advice to that provided him by the AEC and the Pentagon. Unlike "parochial generals" and AEC representatives these scientists appeared to be able to view the nation's problems as detached "objective" observers with a good understanding of the complex issues of military policy. Most importantly of all, in contrast to most of his other top advisors, these scientists shared President Eisenhower's growing concern over the course of the nuclear arms race.

Shortly after the appointment, Killian and the newly elevated PSAC found themselves becoming drawn into the intra-administration struggle over the advisability of a nuclear test ban. Unfortunately for the student of government, however, there are no windows looking in upon these developments of which the scientists quickly became a part. This is the world of closed politics about which C. P. Snow has so eloquently written,[17] and the only available accounts of it are fairly partisan in character.[18] Under these circumstances prudence requires one to be cautious in the assignment of responsibility for decisions which are now criticized in retrospect.

The Intra-Administration Struggle over a Nuclear Test Ban

Although Harold Stassen left the Administration in February 1958, the nuclear test ban issue which he had done so much to raise continued to be debated among Administration officials. On one side were those who favored untying the

[17] C. P. Snow, *Science and Government*, Harvard University Press, 1961.
[18] See, for example, Saville Davis, "Recent Policy Making in the United States Government," *Daedalus (Arms Control* Issue), Vol. 89, No. 4, Fall 1960, pp. 951–66.

American package proposal of August 1957 and then negotiating the nuclear test ban as a separate issue; they included the President's scientific advisors, the Secretary of State,
and to an increasing degree the President himself. On the
other side were the AEC, the Department of Defense, and
their scientific allies in the infinite containment school; by in-
sisting upon negotiating the complete package or nothing at
all, these men in essence opposed a nuclear test ban.

Even though the actual issue at stake was that of the
political desirability of a nuclear test ban, the subject under
discussion was most often whether or not an effective control
system would be technically feasible. The intra-Administration
debate on this and other points was so intense that it spilled
over into newspapers, speeches, and articles; the debate was
brought into the open particularly by the public hearings of
the Senate Disarmament Subcommittee chaired by Senator
Hubert Humphrey.

The leading scientific spokesmen for the conflicting positions
on the wisdom and feasibility of a nuclear test ban were
Edward Teller and Hans Bethe. Teller's main argument
against the test ban was that, given modern scientific technology and the Iron Curtain, the possibilities for evasion would
increase much faster than the possibilities for detection of a
violation of any ban. "In the contest," Teller wrote, "between
the bootlegger and the police, the bootlegger has a great advantage."[19] Such an advantage, Teller argued, would be so
great that it would make any agreement on a nuclear test ban
with the untrustworthy Soviet Union too risky for the United
States.

Hans Bethe, a member of PSAC, took the view that the
risks involved in a nuclear test ban were minor compared to
the political advantages to be gained. And Bethe argued that
the risks, such as they were, could be reduced to a negligible

[19] Edward Teller, "Alternatives for Security," *Foreign Affairs*, Vol. 36,
No. 2, January 1958, p. 204.

point if the control system were extensive enough. Furthermore, Bethe argued, the United States had little to gain militarily from further testing while the Soviet Union through continued testing would be able to come abreast of the United States in nuclear technology.[20]

Prior to the initiation of this latest round in the semi-public debate over a nuclear test ban President Eisenhower had instructed Special Assistant Killian to appoint a special panel to determine the effects on national security of a nuclear test ban and, to a lesser extent, the technical feasibility of a control system to monitor a nuclear test ban. No doubt the panel was also expected to indicate the probable direction of Soviet nuclear developments such as low or high yield weapons and to point out the implications of such an analysis for the desirability of a nuclear test ban.

With the cooperation of the AEC and the Department of Defense, Killian appointed an inter-agency committee under the chairmanship of PSAC member Bethe. Containing representatives from the AEC and the Pentagon, the so-called Bethe Panel included scientists with differing points of view on the questions of the value of continued testing and the feasibility of a nuclear test ban.[21]

In the light of subsequent events, the composition of the Bethe Panel has been criticized in two respects. The first criticism is that the panel was overrepresentative of nuclear physicists and did not contain among its membership any seismologists; these critics further point out that while there were earthquake seismologists on the subcommittees of the panel, there were no explosion seismologists represented even on the

---

[20] These contrasting arguments will be dealt with in detail in Chapter Nine. Bethe's ideas were also touched upon in Chapter Five.

[21] The membership of the Bethe Panel was the following: Dr. Hans Bethe, Chairman; Dr. Harold Brown, Maj. General Richard Coiner USAF, General Herbert Loper, Dr. Carson Mark, Mr. Doyle Northrup, Dr. Herbert Scoville, Jr., Dr. Roderick Spence, Brig. General Alfred Starbird, Col. Lester Woodward USAF, Dr. Herbert York.

subcommittees.[22] The second criticism is that although there was a Department of State observer at the panel's discussions, there were no political experts on the panel itself; as a consequence, these critics argue, the political aspects of the test ban issue were not thoroughly examined.

The defenders of the Bethe Panel dismiss these criticisms on the grounds that the issue under consideration was a technical one primarily within the competence of nuclear weapons experts, i.e., the effect of a test ban, with or without Russian cheating, on American security. Thus they argue that there was no need to include political or seismic experts on the panel. The larger political issue of the desirability of a test ban was, they point out, the responsibility of the National Security Council; the feasibility of detecting underground tests was determined by seismologist advisors to the panel.

Following a period of extensive analysis during which it utilized the services of many scientists the Bethe Panel reported in early April to PSAC, and through it to the President, findings which have been characterized as "unanimous with differences in interpretation." Specifically, the panel concluded that American nuclear weaponry was sufficient in terms of American military requirements and Soviet nuclear capabilities to permit a nuclear test ban without prejudice to American security. Furthermore, the panel decided that while no inspection system to monitor a nuclear test ban could be foolproof, the risks involved therein could be made "acceptable." This is to say that even though the Russians might cheat to a limited extent, the risk to the United States was small.

Although the panel devised a number of possible control systems, the most rigorous system drawn up would have required at least twenty-four inspection stations within the Soviet Union and "highly mobile" inspection teams which

[22] See the discussion of this point in the testimony of F. Gilman Blake before the House Subcommittee on Oceanography, U.S. House of Representatives, Committee on Merchant Marine and Fisheries, Hearings Before a Subcommittee, *Oceanography 1961—Phase 3,* 87th Congress, 1st Session, 1961, pp. 189–90.

could fly to any spot within any nation.[23] While the information has never been made public, the available evidence indicates that the control system proposed by the panel would have had a capability of detecting underground nuclear explosions down to approximately one to two kilotons.[24]

With respect to the feasibility of detecting, the most crucial question for the Bethe Panel to answer had been whether or not a control system would be adequate to monitor underground nuclear tests. Unfortunately there was little empirical knowledge available upon which such a judgment could be based. This situation was well described much later by George Kistiakowsky: "The negotiations on nuclear test cessation have shown the importance of scientific and technological factors for the formulation of national policy in this area. These factors had to be evaluated by *ad hoc* groups that found a *dearth of experimental data* on which to base their conclusions."[25]

The United States had detonated its first underground nuclear explosion just a few months prior to the convening of the Bethe Panel. This shot, code-named Rainier, was equivalent to 1.7 kilotons of TNT and had been suggested by Edward Teller and David Griggs to determine the feasibility of testing underground. It was on the basis of this one underground explosion that the panel sought to evaluate the feasibility of detecting underground explosions.

However, as Kistiakowsky's statement indicates, the lone Rainier shot left unanswered many questions concerning the detectability of underground tests. Edward Teller, for example, was quick to point out that the important problem of possible evasion techniques had been left unresolved by the panel. On the other hand, those scientists who favored

[23] *The New York Times,* April 9, 1958, p. 1.
[24] This statement is based on the initial American presentation at the Geneva Conference of Experts in July 1958 and an article by Thomas Murray, "Nuclear Testing and American Security," *Orbis,* Vol. 4, No. 4, Winter 1961, p. 408.
[25] George Kistiakowsky, "Science and Foreign Affairs," *Bull. Atom. Sci.,* Vol. 16, No. 4, April 1960, p. 115. (Italics mine.)

a nuclear test ban believed further research would serve to strengthen the case for the feasibility of detection. For this and other reasons as well the conclusion of the Bethe Panel that a nuclear test ban was technically feasible permitted a wide latitude of interpretation. Opposed evaluations rested ultimately upon non-technical assumptions concerning the political desirability of a nuclear test ban. It is doubtful, however, that this fact was sufficiently appreciated within the Administration at this time.

While the Bethe Panel had been carrying out its extensive analyses of the technical feasibility of a nuclear test ban the public and intra-Administration debate had continued. During this period two events occurred which strengthened the position of the advocates of a nuclear test ban. The first, rightly or wrongly, discredited Lewis Strauss, Chairman of the AEC and the chief advocate of continued testing; the second strongly reinforced the arguments for a nuclear test ban.

The first episode which was to harm the cause of the advocates of continued testing occurred in early March 1958 and centered on a key question with which the Bethe Panel was working, i.e., the detectability of underground nuclear explosions. In its official announcement of March 6th on the Rainier underground explosion, the AEC had stated that the maximum distance to which the shot had been detected was only 250 miles. This statement brought immediate outcries from many scientists who took their case to the Senate Foreign Relations Committee. After a series of moves by the AEC which only served to increase suspicions that it had deliberately falsified its original report, on March 11, 1958, the AEC revised the maximum distance from 250 miles to 2300 miles. The apparent dishonesty of this maneuver did much to discredit the case made by the opponents of a nuclear test ban that detection was not technically feasible.[26]

A far more serious occurrence which strengthened the posi-

[26] For a discussion of this episode, see James Reston in *The New York Times,* March 12, 1958, p. 1.

tion of the test ban proponents was the Russian announcement on March 31 that it would discontinue its nuclear tests forever provided all other nations ceased testing also. In a world increasingly fearful of radioactive fallout and atomic war, this announcement had a profound effect and increased the pressure on the United States for a test suspension. The timing of the Soviet announcement just after the completion of its own test series and just before an announced United States series was particularly excellent in terms of its propaganda impact.

Despite official comments to the contrary, this propaganda defeat had a profound effect on Administration thinking. And although Secretary of State Dulles announced that the United States had considered and rejected a similar move because the Administration favored the continued development of "clean nuclear weapons for use in air defense and ground warfare," it is evident that the Russian propaganda victory had a significant impact on the thinking of the President and his Secretary of State.

In response to these events and influenced by the views of Killian and the finite containment scientists on PSAC, Secretary of State Dulles concluded that the United States must make a disarmament proposal with a good probability of success.[27] Secretary Dulles therefore suggested to the President the idea of technical talks between East and West scientists on a system to police a nuclear test ban and on means to prevent surprise attack. The President agreed with Dulles and asked him to draft a letter to the Soviet Premier incorporating Dulles' ideas.

In the resultant letter to Khrushchev of April 28, 1958, President Eisenhower reiterated the suggestion he had made in January 1958 that East and West convene technical talks on

[27] This discussion is based on Saville Davis' "Recent Policy Making in the United States Government," *op.cit.*, pp. 961–62. Informants indicate that Davis' account is substantially correct with respect to the facts of this matter.

the inspection systems necessary to supervise any forthcoming disarmament agreements. The President's new proposal for technical talks contained, however, one important modification of previous proposals. Whereas the call for technical talks in the past had been part of a larger package proposal, the President seemed to be suggesting for the first time that the United States would agree to a nuclear test ban as an isolated step provided the Soviet Union would agree to a workable inspection system. And, in subsequent conferences with the press, Secretary of State Dulles publicly supported this view that the United States was breaking up its package proposal of August 1957 and might be willing to negotiate a nuclear test ban as a first step arms control agreement.

The favorable Russian reply to the President's suggestion appeared to indicate a similarly important change in Russian policy on disarmament. For the first time, the Soviet Union seemed to be ready to accept the necessity of an inspection system to verify the carrying out of arms limitation agreements. If so, a major breakthrough in the disarmament negotiations was in the offing. The Western hope for such a development was reinforced when, after further exchanges, the USSR agreed to the convening in Geneva on July 1, 1958, of technical discussions on methods to police a possible nuclear test ban. In addition, it was subsequently agreed that technical talks on the prevention of surprise attack be convened in November 1958.

Nevertheless, despite its concurrence on the necessity for technical talks to determine the technical feasibility of a nuclear test ban control system, the Soviet Union continued to press for a total ban on further testing. Through public statements, the Soviet Union sought to create the impression that the technical talks were intended to be preliminary to a treaty banning future nuclear tests rather than merely an effort to determine whether or not a control system was feasible if the nuclear powers should ever actually reach an agreement on a test ban. This attempt of the Russians to have the United States commit itself to a total ban on further testing reached its high point when the USSR threatened to boycott the technical talks unless the United States renounced further testing.

The United States, on the other hand, insisted that the technical talks were exploratory only and were to be entered into without commitment by either party. President Eisenhower, in his letter to Chairman Khrushchev confirming the technical talks, made the United States position quite clear that the "talks would be undertaken without commitment as to the final decision on the relationship of nuclear test suspension to other more important disarmament measures I have proposed."[28]

Emphasizing its position that the technical talks did not constitute a commitment to a nuclear test ban, the United States was quite adamant in its insistence that the title of the conference reflect this condition. The conference was to be known officially as the "Conference of Experts to Study the Possibility of Detecting Violations of a *Possible* Agreement on Suspension of Nuclear Tests" (italics mine). With this understanding in mind, the American scientific delegation, composed of James Fisk (chairman), Robert Bacher, and E. O. Lawrence, met with their Russian and British counterparts at Geneva on July 1, 1958.[29] However, as the talks were soon to prove, the relationship between the technical and the political is far more subtle and intricate than the American scientists or political leaders realized at the time. Of this more will be said in Chapter Seven.

[28] Conference of Experts to Study the Possibility of Detecting Violations of a Possible Agreement on Suspension of Nuclear Tests, *Verbatim Transcript of Second Meeting*, July 2, 1958, pp. 27–30.

[29] Fisk and Bacher were both members of PSAC and of the finite containment school. Fisk is President of Bell Laboratories and Bacher is Professor of Physics at the California Institute of Technology. Lawrence, a member of the infinite containment school, was Director of the Livermore Weapons Laboratory. In addition there were numerous scientific advisors and observers from interested government agencies. Lawrence was to withdraw from the Conference shortly after its commencement due to ill health; he died shortly thereafter.

The Geneva System

To scientists familiar with the Geneva System and its scientific foundations the succeeding technical discussion will undoubtedly appear to be oversimplified. This oversimplification is necessary in order to make a very complex subject more understandable to those less familiar with it. Hopefully, however, the writer has been able to maintain the essence of the subject and has not distorted any important considerations.

THE TECHNICAL PROBLEM FACING THE CONFERENCE OF EXPERTS

The system to police a nuclear test ban drawn up by the Conference of Experts and called the Geneva System was based on the fact that the products of nuclear explosions such as radioactive debris, electromagnetic waves, acoustic waves, and seismic waves can be detected by surveillance devices stationed about the earth and in space. The type of device which will detect an explosion is dependent upon the medium in which the explosion takes place, i.e., in the atmosphere, in the ocean, in the earth, or in space.

Since, however, many natural events in each medium produce disturbances very similar to those produced by nuclear explosions, the crucial problem is to be able to *identify* a detected event as a nuclear explosion. The difficulty of achieving such identification differs for each medium. In the case of atmospheric explosions identification can be positive through the examination of radioactive debris surrounding the test area or in clouds originating at the test site. Thus, if a suspicious atmospheric event were detected through electromagnetic waves, it would be relatively easy to determine whether or not the event had been a nuclear explosion by finding nuclear debris at the scene of the event. For test shots in the world's oceans, *detection* is relatively easy and certain, but acquisition of radioactive debris, the final proof of *identification,* is much more difficult.

In the case of underground tests where there is no radio-activity vented into the atmosphere, the only observable signs through which such explosions can be detected are the earth or seismic waves set in motion by the explosion and registered on the world's seismographs. However, at the time of the Conference of Experts and of this writing, there was and is no positive way to identify an underground nuclear explosion merely through a study of the properties of the seismic signal. This is so because, while most earthquakes and earth tremors may be identified as such by the seismic reading, a sizeable number of earthquakes annually give the same type of recording on a seismograph as do underground explosions. As a result, frequently the only way in which it could be determined whether or not a particular event registered by a seismograph were a nuclear explosion would be through examination by drilling of the earth where the event occurred; then the discovery of radioactive debris would positively identify the detected event as a nuclear explosion. If no debris were discovered it could be assumed that the event had been an earthquake. However, unless there were surface manifestations such as fissures in the earth there would be no positive way to identify the detected event as an earthquake.

In contrast to nuclear explosions in the other three media, nuclear explosions at high altitudes and in space give rise to *no* means of *positive identification*. In fact, nuclear explosions in these media are presently difficult, if not impossible, even to *detect*. The twin problems, therefore, are to devise methods by which the radiation from a nuclear explosion in space or at high altitudes may be detected and to determine criteria which would allow an inference to be made that the detected phenomena such as electromagnetic waves, gamma rays, neutrons, and X-rays were produced by a nuclear explosion rather than by some natural event such as a solar flare.

The task, then, facing the technical negotiators in the summer of 1958 was the devising of methods and a network of control posts which would detect and, if possible, identify all events in all media which might be nuclear explosions. As a network of control posts which is large enough to detect underground explosions presumably would be one that is large enough to detect explosions in other media as well, the governing decisions in devising a control system relate mainly to the problem of detecting *underground* nuclear explosions.[30] For this reason, the primary emphasis in this discussion will be on the detection and identification of these underground explosions.

The fact that there is no positive way to identify underground nuclear explosions through a study of resulting seismic waves means that the problem of identification has to be approached indirectly. The requirement for the control system is to identify through seismic readings a high percentage of the detected events as natural phenomena and to leave the small number of remaining suspicious events to be identified through on-site inspections, i.e., search for radioactive debris by drilling.

At the time of the Conference of Experts it was believed that the identification by seismic readings of a high proportion of detected underground events as natural earthquakes was possible through application of what is known as the criterion of the "first motion." On the basis of theoretical studies and the "Rainier" nuclear test shot, it was assumed by the Conference of Experts that all detected underground nuclear explosions would register a *positive* first motion on a seismograph. This is to say that the explosion would push away the earth in *all* directions and cause the pendulum on all detecting seismographs to swing upward or positively. Earthquakes, on the other hand, being due to a slippage in the earth's crust, may cause downward or negative first motions on detecting seismographs located in some directions from the earthquake and positive first motions on seismographs in other directions

---

[30] The exception to this is that explosions at very high altitudes and in space would be detected primarily by the use of earth satellites and not solely by use of control stations on the earth's surface.

from the earthquake. A number of earthquakes, however, like underground explosions, do give rise only to positive first motions in all directions.

If, then, a control post should record a negative first motion in conjunction with a seismic event, that event could be identified automatically as an earthquake. If, on the other hand, no control post recorded a negative first motion while some control posts did register positive first motions, the event would be regarded as suspicious. It might have been caused by a nuclear explosion or else by an earthquake which produced no *detected* negative first motion. On-site inspection would then be required to determine which had been the case.

The probability that the first motion of a seismic event will be detected by a control post is dependent in general upon the size or yield of the event[31] and the distance of the event from surrounding control posts.[32] As the yield of the event increases, the probability that the first motion will be detected increases correspondingly. Similarly, as the number of control posts in the network increases and, as a consequence, the distance from the control posts to seismic events decreases, the probability that a first motion will be detected increases. Thus, as the yield of the event and the number of control posts increase, the probability will increase that suspicious events will be *detected* by their *positive* first motions and that earthquakes will be *identified* by their *negative* first motions. On the other hand, due to the fact that the number of earthquakes equivalent to a given yield explosion increases as the yield of the detected event decreases, the number of unidentified events requiring on-site inspection for identification purposes would increase as the yield of the nuclear explosion to be monitored by the system decreases.

On the basis of these relationships, the Conference of

[31] "Size" and "yield" mean the energy of the earthquake or nuclear explosion.
[32] This generalization overlooks many complicating factors such as the skip distance, interference, seismic "noise," and decoupling (methods by which to reduce the resultant seismic wave—see Chapter Eight).

Experts fashioned a control system for presentation to their respective governments. This network of control posts which has become known as the Geneva System was determined by the Conference of Experts to have the characteristics, capabilities, and limitations discussed below.

THE CHARACTERISTICS OF THE GENEVA SYSTEM

The system proposed by the Conference of Experts was to be composed of 160 to 170 land-based control posts and about ten ships. About 100 to 110 of the land-based control posts would be situated on continents, twenty on large oceanic islands, and forty on small oceanic islands. However, while it was decided that there should be thirty-seven control posts located in Asia, the exact location was left undecided. Nor was it decided how many posts would be located within a particular nation, such as the Soviet Union. It was apparently assumed by the experts that this could be settled by subsequent technical talks.

Two other aspects of the Geneva System must also be mentioned. Firstly, the experts agreed that "when it is necessary to investigate whether a radioactive cloud is present, in the case of detection of an unidentified event which could be suspected of being a nuclear explosion, special aircraft flights would be organized in order to collect samples of radioactive debris. . . . "[33] However, the conditions under which such flights could be initiated over the territory of the participating nations were left unspecified by the experts.

Secondly, the experts, noting that the task of identifying underground, oceanic, and low altitude explosions depended upon the employment of inspection methods to discover telltale radioactive debris, made provision for the carrying out of

---

[33] U.S. Congress, Joint Committee on Atomic Energy, *Hearings on Technical Aspects of Detection and Inspection Controls of a Nuclear Weapons Test Ban*, 86th Congress, 2nd Session, 1960, pp. 523–24.

These flights would be in addition to regular flights carried out by the control organization over the oceans to sample the atmosphere for evidence of nuclear debris.

on-site inspections by the control commission. In the event of a suspicious event, " . . . the international control organ," their report read, *"can send* an inspection group to the site of this event in order to determine whether a nuclear explosion had taken place or not."[34] Again, however, the details concerning the initiation of an on-site inspection were left unresolved.

THE CAPABILITIES OF THE GENEVA SYSTEM

The capabilities of the Geneva System, as judged by the Conference of Experts, differed for the various media, such as the atmosphere, the oceans, and underground. For this reason, plus the fact that the experts' evaluations are expressed in qualitative terms, such as "good probability," it is difficult if not impossible to judge the overall capabilities of the system drawn up by the experts. Whether it is even possible to make any valid determinations of the capabilities of the system is a controversial matter which will be discussed in Chapter Nine.

With respect to atmospheric nuclear explosions near the surface of the earth (below ten kilometers), the experts believed the Geneva System would have a "good probability of detecting and identifying nuclear explosions of yields down to about [one] kiloton." For explosions taking place at altitudes from ten to fifty kilometers, the system would have a "good probability of detecting, but not always of identifying, [nuclear] explosions . . . "[35]

As for nuclear explosions taking place in the world's oceans, the experts believed the system would have a "good probability of detecting nuclear explosions of [one] kiloton yield. . . . " However, they pointed out that "the identification of underwater explosions can, in comparatively rare cases, be made more difficult by natural events which give similar hydroacoustic and seismic signals."[36]

---

[34] *ibid.*, p. 524. (Italics mine.)
[35] *ibid.*
[36] *ibid.*

Due to the large numbers of unknowns, the experts could conclude only that it was *theoretically possible to detect* nuclear explosions at high altitudes (above fifty kilometers) and in space. However, they did mention in their report the possibility that shielding could be used to absorb telltale gamma radiation from such an explosion and thereby reduce the possibility of detection.

On the other hand, with respect to underground explosions, the Geneva System, in the judgment of the experts, had a capability, or probability, of *identifying as earthquakes* 90 per cent of all seismic events with a yield of five kilotons or more; this would leave a remaining 10 per cent of detected seismic events with a five kiloton yield or above which would require an on-site inspection to identify their nature. The experts estimated that the number of such unidentified and, therefore, presumably inspectable events would be between 20 and 100 annually. The number 20 was supplied by the Soviet Union; the United States scientists believed the number would be closer to 100.

While the Geneva System would record underground events with a yield of less than five kilotons, the percentage of these events that could be positively identified as earthquakes by their seismic readings would drop off very rapidly as the yield of the seismic event decreased. As a consequence the number of suspicious events recorded below five kiloton yield necessitating on-site inspections for identification would rise very rapidly. However, the experts did not attempt to predict in their report what the number of required inspections or the probability of identification might be at any yield less than five kilotons.

It was the belief of the experts that "although the control system would have great difficulty in obtaining positive identification of a carefully concealed deep underground nuclear explosion, *there would always be a possibility of detection* of such a violation by inspection."[37] The experts did not, how-

[37] *ibid.*, p. 525. (Italics mine.)

ever, clarify the meaning of this extremely ambiguous statement.

In summary, although the Geneva System, in the judgment of the experts, promised a "good probability" of detecting and identifying nuclear explosions of one kiloton TNT equivalent and above at the surface of the earth, in no other media could it promise an equal performance. Its ability to identify underground nuclear explosions was uncertain at the least and quite limited at the most. As for the detection and identification of nuclear explosions in space and at high altitudes its capabilities were almost nonexistent.

THE ASSUMED LIMITATIONS OF THE GENEVA SYSTEM

In discussing the capabilities of the Geneva System as envisioned by its creators, we have already mentioned two major limitations by means of which a nation could evade the system. The first is the possibility of reducing detection and identification of explosions in space through the use of shielding; the second is the difficulty of identifying underground explosions of small yield, i.e., under five kilotons. In addition to these limitations of the system, the report of the experts pointed out the possibility for additional evasions: " . . . in certain special cases the capability of detecting nuclear explosions would be reduced; for instance, when explosions are set off in those areas of the ocean where the number of control posts is small and the meteorological conditions are unfavorable; in the case of shallow underground explosions; when explosions are set off on islands in seismic regions; and in some other cases when the explosion is carefully concealed. In some cases it would be impossible to determine exactly the area in which a nuclear explosion that had been detected took place."[38]

The experts believed, however, that these limitations did not seriously impair the efficacy of the control system. "Whatever the precautionary measures adopted by a violator," the experts concluded, *"he could not be guaranteed against*

[38] *ibid.*, p. 526.

exposure, particularly if account is taken of the carrying out of inspection at the site of the suspected explosion."[39] Unfortunately, the meaning of this important expression was left unexplained in the experts' final report.

The Political Impact of the Experts' Report

On August 21, 1958, after seven weeks of amicable negotiations, the scientist-diplomats released their final communique and completed their "technical" report. The Western delegation was especially pleased because, despite the intransigence of the Russian scientists on certain points, it believed it had succeeded in obtaining technical agreement on a control system to monitor a nuclear test ban. For the first time—or so it appeared—the Russians had agreed to the principle of international inspection. The scientists thus had successfully fashioned a technical solution to one aspect of the atomic weapons problem, a solution based on the "logic of the facts." "As scientists," Fisk told the Conference, "we have sought here to establish the facts pertinent to our subject, and to draw from them sound and logical conclusions regarding a system of control."[40]

The final communique drafted by the Soviet delegation and approved by the Western delegations reflected this mood of optimism: "The Conference came to the conclusion that the methods of detecting nuclear explosions available at the present time . . . make it possible, within certain specific limits, to detect and identify nuclear explosions, and it recommends the use of these methods in a control system."[41]

This mood of optimism spread rapidly throughout the world. As far as the peoples of the world were concerned, the experts had proved that a 180 control station net would detect *all* nuclear tests *everywhere* (except very small explo-

[39] *ibid.* (Italics mine.)
[40] *The New York Times,* August 22, 1958, p. 4.
[41] Conference of Experts . . . , *op.cit., Verbatim Transcript of the Thirtieth Meeting,* August 21, 1958, pp. 4–6.

sions)[42] and thus the world believed the last major roadblock to a nuclear test ban had been removed. As the late Secretary-General of the United Nations, Dag Hammarskjold, put it, the experts had made "an effective dent in the hitherto rather intractable problem of disarmament. It will hereafter lie with the Governments concerned and the United Nations to follow through the opening you have created. I have every hope that in due course they will indeed make the necessary constructive effort."[43]

In the United States, President Eisenhower too was pleased that at last a concrete first step might be taken to remove the deadly threat of nuclear weapons. "Any step like this," the President told his press conference on August 21, 1958, "that proves that you have a real agreement between intelligent people of both sides, gives grounds to hope that you can go another step, and every step that you go means you can go another one."[44]

The next day, August 22, 1958, President Eisenhower, on the recommendation of Secretary of State Dulles, proposed that the three nuclear powers negotiate a treaty for the permanent suspension of nuclear weapons testing. In addition, the President announced that the United States would suspend its nuclear tests for one year and would extend the suspension beyond that if three conditions were being met: (1) if all other nations suspended their tests; (2) if an effective inspection system were established to police the ban; and (3) if "satisfactory progress is being made in reaching agreement on and implementing major and substantial arms control measures such as the United States has long sought."[45] On the basis of this United States initiative, it was subsequently agreed by East and West that negotiations to imple-

---

[42] See, for example, the coverage in *The New York Times,* August 22, 1958.

[43] Conference of Experts . . . , *op.cit., Verbatim Transcript of the Thirtieth Meeting,* August 21, 1958, p. 7.

[44] *The New York Times, August* 22, 1958, p. 12.

[45] *The New York Times,* August 23, 1958, p. 1.

ment the Geneva System would convene on October 31, 1958, in Geneva.

Conclusion

President Eisenhower's proposal to negotiate a nuclear test ban agreement as an isolated step was in effect a major reversal of American nuclear policy. While it can be characterized as a conditional acceptance of the first step philosophy advocated by his scientific advisors, it would be a mistake to place primary responsibility for the shift in the Administration's position upon these advisors.[46] It is even more misleading to suggest, as some have, that the United States decision to seek a nuclear test ban was due to the fact that Killian "favored scientific opinions of those who are considered extremely naive about communism, such as Dr. Hans Bethe of Cornell University."[47] For while one may disagree with Bethe and his colleagues on the wisdom of a nuclear test ban, their selfless efforts to insure the strength of the Free World belie the charge that they are naive about the seriousness of the Communist threat. Furthermore, it was Secretary Dulles who recommended to President Eisenhower the advisability of negotiating a nuclear test ban.

Most importantly, it appears that the President's desire for a nuclear test ban had become so strong by the fall of 1958 that it overrode both his previous position against a nuclear test ban as an isolated step and his conviction (stated as late as April 9, 1958) that the United States should not suspend testing until it had learned all it desired to learn about nuclear weapons—or at least the nuclear weapons then being developed.[48] That this last condition had not been met

[46] For example, the late Thomas Murray did this when he implied that PSAC had misled the President on the technical feasibility of a nuclear test ban. "Nuclear Testing and American Security," *Orbis*, Vol. 4, No. 4, Winter 1961, pp. 405–21.

[47] This quotation comes from an editorial of the *Manchester* (NH) *Union Leader* which was inserted in the *Congressional Record*, July 20, 1961, p. A5572 by Congressman Gordon Scherer.

[48] See Presidential Press Conference as reported in *The New York Times*, April 10, 1958, p. 1.

is evident from the fact that it was *prior* to the completion of the American test series then in progress when the President proposed the political conference to draw up a treaty for a permanent test suspension. Furthermore, the President called for political negotiations on a nuclear test ban despite the fact that, according to a statement by Secretary Dulles, he had rejected this idea the previous spring in favor of continued research on a "clean" bomb; only the summer before, Teller and Lawrence had said "four or five years [were needed to] . . . produce an absolutely clean bomb."[49]

This is not to suggest, however, that the United States entered into the test ban negotiations prior to meeting its then projected military requirements. On the contrary, the available evidence suggests that the test series in the fall of 1958 enabled the United States to prove out its latest weapons designs such as improved missile warheads and tactical weapons.[50] Whether or not these weapons would meet future American military requirements was, of course, a central question in the debate over the advisability of a nuclear test ban. The victory for the pro-ban side in this debate was due to President Eisenhower's shift to their position that the political gain to be derived from a nuclear test ban was greater than the possible military advances the future might hold.

President Eisenhower's own attitude on disarmament has been revealed by his former political chief of staff, Sherman Adams. "The President," Adams writes, "spent long hours of discussion in search of ideas to break the disarmament deadlock with Russia." At the end of one particularly frustrating session with his disarmament advisors the President, Adams reports, closed the meeting with the following ejaculation: "Something has got to be done. We cannot just drift along or give up. This is a question of survival and we must put

[49] *The New York Times,* June 27, 1957, p. 1.

[50] A statement by John McCone bears out this contention. A summary of the available evidence is contained in George Herman's "To Test or Not to Test," *The New Leader,* November 13, 1961, pp. 6–8.

our minds at it until we can find some way of making progress. Now that's all there is to it."[51] Such an attitude made the President responsive to any ray of hope for a breakthrough in the long deadlocked negotiations.

Of equal importance as a cause of the reversal in American policy was the apparent change in the Soviet attitude. In spite of numerous setbacks in the disarmament negotiations there was reason for the President and his Secretary of State to believe that since early 1957 the Russians had become genuinely interested in limiting the arms race. The President may have reasoned that the Communists too were disturbed by the economic burden caused by weaponry and were realizing as well the dangers inherent in the nuclear arms race.

Thus it appears that the great desire for progress in disarmament and the growing belief that negotiations might be fruitful underlay the change in the American position of a nuclear test ban. The scientists at Geneva appeared to have made a breakthrough and thus to have imposed upon the Administration a responsibility to seize the opportunity which had been created. Such considerations and not an unwarranted influence by the President's scientific advisors appear to account for the culmination in the shift in American disarmament policy which took place on August 22, 1958.

On the other hand, it is no doubt justifiable to argue, as does Saville Davis, that the President's scientific advisors exerted a strong influence on his decision. Whether or not the absence of any infinite containment scientists as immediate advisors to the President made it easier for him to make the decision he did is impossible to answer. If they had been in a position to press their technical arguments against a test ban they *might* have been better able to make him aware of the weaknesses in the Geneva System as they saw them. It is difficult to believe, however, that the scientific aspects of the issue were such that the presence of strong anti-ban scien-

[51] Sherman Adams, *First Hand Report—The Story of the Eisenhower Administration,* Harpers, 1961, pp. 318–26.

tists among the President's immediate advisors could have changed his mind.

The President was determined to find a way out of the nuclear arms race and a number of factors made easier for him the decision he very much wanted to make. In the first place, the replacement of Lewis Strauss by John McCone as chairman of the AEC (July 1, 1958) removed from the policy debate a strong opponent of a nuclear test ban. In the second place, the military were in some disfavor at the White House because of their "parochial" approach to national security problems. And, in the third place, the Joint Committee on Atomic Energy was not informed of the President's intention to call for the political negotiations until August 21, 1958 and by then it was too late to change the President's course of action.

The change in American nuclear policy can be accounted for by the confluence of many factors, only one of which was the influence on the President of the finite containment scientists. Nevertheless, the fact that there were no infinite containment scientists on the President's Science Advisory Committee caused these men to believe, as they also had believed in 1949 with respect to the H-bomb issue, that their views had not been properly presented to the President by his scientific advisors. As a consequence they believed they had been dealt an injustice.[52]

The most significant lesson to be drawn from this history of the decision of the United States to seek a nuclear test ban is the need to avoid in the future any similar failure to appreciate the intertwining of the technical and political aspects of such issues. All the participants in making that decision believed

[52] As was mentioned earlier, Lawrence, the representative of the infinite containment group on the U.S. delegation, left Geneva in mid-July due to ill health. He was not replaced although other infinite containment scientists remained as advisors. Whether or not Lawrence's departure affected the course of the negotiations, as some argue, is difficult to judge. It is fairly common knowledge, however, that Lawrence was sent to Geneva at the insistence of Lewis Strauss who wanted someone in the delegation sympathetic to his own position.

that a strict separation could be maintained between the technical and political realms; furthermore there was no thorough examination of the test ban issue by a panel of political, scientific, and military experts. If an examination had been made by such experts rather than nearly total reliance being placed upon the Bethe Panel, PSAC, and the President's immediate political advisors, the Administration might have developed a policy which could have avoided subsequent errors in the test ban negotiations.

The consequences of the belief that the technical and the political realms could be isolated from one another are most evident in the American participation in the Geneva Conference of Experts. This failing on the part of the Administration, of the American scientific delegation to the conference, and of the scientific community in general was to be costly for national policy and the unity of the scientific community. Because this matter has an intrinsic interest to the student of national policy-making in this scientific age and also has had a profound effect on American nuclear policy, this study now turns to an evaluation of the Geneva Conference of Experts and of the reactions of scientists to the conference.

CHAPTER SEVEN

THE CONFERENCE OF EXPERTS: AN EVALUATION AND THE SCIENTIFIC REACTION

IN PRIVATELY expressed opinions and to some extent in print, the performance of the American scientific delegation at the Geneva Conference of Experts has received considerable criticism. It is with reluctance that the writer enters into this controversy which has already aroused the passions of those involved; furthermore, he deems it essential to make a number of qualifications prior to an evaluation of the Geneva Conference of Experts and to a discussion of the reactions of scientists to the outcome of the conference.

Firstly, discussions of the Conference of Experts must be undertaken without access to all of the relevant facts. While this limitation is to some degree true with respect to almost all historical problems, the missing pieces in the present history could be particularly important in the evaluation of the events which took place. For example, it could well be that they who argue that Secretary of State Dulles agreed to negotiate a nuclear test ban treaty principally to call the Russians' bluff on inspection are correct, and that everything that has happened is explicable in light of this fact.

Secondly, the American scientists went to Geneva with knowledge of President Eisenhower's intense desire for a nuclear test ban. The importance attached to the technical talks by the President can be judged by the fact that he established an *ad hoc* Committee of Principals outside the National Security Council to advise him on a nuclear test ban. The members of the Committee were the following: the Secretary of State; the Special Assistant for Science and Technology; the Chairman of the AEC; the Director of the Central Intelligence Agency (CIA); and a representative of the Secretary of Defense.

And, lastly, the inability of the Committee of Principals to reach a unified position on a nuclear test ban created a policy vacuum within which the "technical" negotiations had to operate. The President, the Secretary of State, the head of the CIA, and the Science Advisor favored a nuclear test ban; the Chairman of the AEC and the Pentagon were opposed. Failing to agree on what they wanted, the Committee members could not effectively or intelligently direct and evaluate the activities of the American experts at Geneva. For this reason the American scientists at Geneva apparently were left largely on their own in the conduct of the negotiations.

Appraisal of the Geneva Conference of Experts

THE DIPLOMATIC RESPONSIBILITY OF THE SCIENTISTS

In retrospect there can be no doubt that the Conference of Experts was unique in the annals of diplomacy. For, while the responsibility of the scientists was ostensibly solely technical in nature, in reality these men had been assigned a political task of the highest order. On a matter of grave consequence the Great Powers—or as it now appears, at least the United States—had entrusted to a group of untrained private citizens the serious diplomatic responsibility of negotiating the broad outlines of an arms control agreement. The significance of this fact, however, had not been recognized in early July 1958, and as a consequence the American experts were soon to find themselves in a position for which they were ill-prepared. Perhaps, therefore, their strengths at the diplomatic bargaining table should be more impressive than their weaknesses. At the least, the following criticisms relate to the lack of foresight and experience of the scientist-diplomats and not to any lack of integrity or intelligence.

The American scientists went to Geneva with the belief that although their task had great political significance, their responsibility was solely technical, i.e., to assess only the technical possibilities of policing a possible agreement to ban nuclear weapons tests; this was made quite explicit by James Fisk in his opening statement to the Conference. "We hope," Fisk told the assembled delegations, "that this inquiry can be kept exclusively technical. Our side is not empowered to discuss or reach decisions on any political matters. It will ease our deliberations if we are able to confine our discussions during the course of these talks to the technical issues we face."[1]

The United States delegation initially took the position that, after a consideration of all the technical factors, the conference should devise *"a number of systems"* with varying characteristics, capabilities, and limitations which would then be presented for consideration to their respective governments. Specifically, the conference was to draw up a "family of curves"; on one axis there would be listed a spectrum of control systems ranging from relatively few to many control posts; on the other axis there would be listed the estimated identification-detection capabilities of each specified control system. Then the responsibility for the selection of a particular system for implementation would rest with the governments at subsequent political conferences.

In decided contrast, the Russian experts from the very first day of the conference argued that the purpose of the conference was to prepare *"the entire control system* for observing the agreement on the cessation of nuclear tests."[2] The experts, according to the Soviet scientists, should look into the various "existing methods of detecting nuclear explosions and [should select] . . . from among them those which could be recommended to the governments as the most effective . . . ."[3] The

---

[1] Conference of Experts to Study the Possibility of Detecting Violations of a Possible Agreement on Suspension of Nuclear Tests, *Verbatim Transcript of First Meeting,* July 1, 1958, p. 12.

[2] *ibid., Verbatim Transcript of Third Meeting,* July 3, 1958, pp. 6–10. (Italics mine.)

[3] *ibid.,* pp. 3–5.

Soviet scientists sought to maneuver their Western colleagues into an implied commitment to a test ban and to *a* particular system by which to police the test ban.

On a second matter as well, the Western delegations were in fundamental political disagreement with their Russian colleagues. Not only did the Russians insist that the conference agree on one system which would be proposed to their respective governments, but they had definite ideas on the *political* requirements for such a system. For example, whereas the Western delegations wanted the conference to draw up specifications for a system with a capability to detect underground nuclear explosions down to one kiloton TNT equivalent in all environments, the Russian scientists refused to consider such a system. The cause of this Russian intransigence was explained two years later to the Joint Committee on Atomic Energy by Harold Brown, a scientific advisor to the American delegation: "[To monitor underground explosions down to] 1 kiloton equivalent or more required, it turned out, something like 650 stations worldwide. However, 650 stations were something that the Eastern delegation would not accept."[4]

Thus the scientific delegations at Geneva found themselves with opposed political instructions. As a consequence, the American experts had to decide whether to break off the negotiations or to attempt to work out a technically sound agreement within the political restrictions imposed upon the conference by the Russian delegation. In choosing to follow the latter course of action, with at least the tacit approval of the Eisenhower Administration, the American experts undertook a task beyond their political experience and competence.

It was this failure of the American scientists to appreciate their political role rather than a departure from scientific objectivity or from professional integrity which was the cause

[4] U.S. Congress, Joint Committee on Atomic Energy, *Hearings on Technical Aspects of Detection and Inspection Controls of a Nuclear Weapons Test Ban,* 86th Congress, 2nd Session, 1960, p. 19.

of the political mistakes they committed at Geneva. Given the state of scientific knowledge at the time, the technical correctness of the experts' report is beyond question. What may be criticized, however, is the quality of the political judgment they exercised in deciding a number of issues which, although in most cases technical in appearance, were really political in nature. Therefore, while the present chapter is critical of the scientists as diplomats, it agrees fully with the frank appraisal by James Fisk of the performance of his delegation at Geneva: "There were no compromises in the technical conference that preceded the political conference. The foundations, technical and scientific, which were then available to the political conference were those based on the best technical evidence which we then had.

"As we know, subsequent information showed that we had been somewhat too optimistic. *I feel that we perhaps should have at that time developed a program to support our political negotiators which might have put us in a better position today than we are in.*"[5]

The optimism of these untrained and amateur diplomats at the close of the conference was understandable. For the first time in the history of postwar disarmament negotiations, the Soviet Union had accepted in principle—or so it seemed—the notion that inspection had to be an integral part of any disarmament agreement. Furthermore, the American scientists no doubt believed that the compromise control system represented a victory for the West; it had a one kiloton capability for tests in the atmosphere or in the oceans while its underground capability extended as low as five kilotons. In addition, the American scientists believed that they had made quite clear the limitations of the control system. And, finally, they had found their Russian counterparts, despite some displays of

[5] U.S. Senate, Committee on Government Operations, *Hearings Before the Subcommittee on National Policy Machinery on Organizing for National Security*, 86th Congress, 2nd Session, Part II, 1960, p. 310. (Italics mine.)

intransigence, to be quite amenable. The belief that scientists could work together where diplomats had failed no doubt gave the American scientists confidence that issues left unsettled by the conference could be clarified at subsequent technical talks. The important fact in the fall of 1958 was that a breakthrough toward an arms control agreement seemed to have been made.

The strong feeling of the American experts, then, that they had a responsibility to reach agreement if one consistent with American interest were at all possible and their apparent high degree of success seem to account for the optimism of the scientists. Unfortunately, this sense of optimism and their confidence in the technical integrity of their report allowed the American scientists to be misled in their capacity as political negotiators. With these considerations in mind let us turn to the proceedings of the conference itself.

THE PERFORMANCE OF THE SCIENTISTS AS DIPLOMATS

Following a detailed discussion among the delegates of the capabilities and limitations of the various methods to detect nuclear explosions, the three delegations each presented a control system incorporating these methods.[6] In accordance with his original statements on the purpose of the conference, Fisk argued that the experts should include all these proposed systems as well as any others which might be suggested in their final report. The Russians, on the other hand, con-

---

[6] The systems proposed by the three delegations and their estimated characteristics are listed below. See Conference of Experts . . . . , op.cit., *Verbatim Transcript of Twenty-Sixth Meeting,* August 5, 1958, pp. 27–30.

	Soviet			British			U.S.		
Number of Control Posts	110			170			650		
Yield (Kt)	1	5	20	1	5	20	1	5	20
Probability of Identification of Earthquakes on Land Areas	Less Than 5%	50%	90%	90%			90%	Greater Than 90%	

tinued to insist that only one system be recommended in the experts' report.

In order to break this deadlock, the Western delegations, in the words of Harold Brown, "made the proposal that the number of . . . [control posts] could be adjusted depending upon what capability or threshold one wanted to give the system."[7] Specifically, the Western delegation suggested the concept of a detection-identification threshold by which the lower limits of the capabilities of the control system would be specified. In this way the experts could proceed to reach agreement on a control system whose number of control posts would be agreeable to the Russians and yet be technically correct.

It was this decision of the Western delegations, presumably reached with the concurrence of their respective governments, that changed their diplomatic task. Whereas their diplomatic responsibility originally had been narrowly prescribed, i.e., to reach a technical agreement with the Eastern delegations on a spectrum of control systems, now their task became that of bargaining with the Communist negotiators and committing the latter to *a* control system with a threshold as close as possible to that of the original American system drawn up by the Bethe Panel.

It is clear from the transcript of the conference that this shift in the diplomatic role of the scientists and in the purpose of the conference had occurred by the twenty-sixth meeting. At that session Fisk noted that, although it might be valuable to present in the final report descriptions of all the systems discussed by the experts, the conference was not limited to this responsibility.[8] He then told the conference that his delegation was willing to submit to its government the 180 station control system proposed by the British in order *"to*

[7] U.S. Congress, Joint Committee on Atomic Energy, *Hearings on Technical Aspects* . . . . , *op.cit.,* p. 19.

[8] Conference of Experts . . . . , *op.cit., Verbatim Transcript of Twenty-Sixth Meeting,* August 5, 1958, p. 61.

obtain an agreed conclusion on the possibility of detecting violations of a possible agreement" to ban nuclear weapons tests.[9]

The reason why the 180 station control system was chosen was explained later by Harold Brown to the Joint Committee on Atomic Energy: "The Western delegation felt that there was a very large number of possible systems of whose capability it was prepared to estimate, with various capabilities for identifying earthquakes at 5 kilotons, or 1 kiloton, or wherever. However, having finally agreed to discuss the capabilities of 180 station system, the Soviets refused to consider any other."[10]

Although he was willing to submit to his government the 180 station system as an agreed conclusion, Fisk continued to propose that the final report include descriptions of all three systems discussed by the scientists. In addition, he reminded the delegates that "we are not recommending a system to our governments, we are transmitting conclusions." In spite of this admonition, however, the final report of the conference failed to include descriptions of all three systems, and it did make a specific recommendation: "The Conference of Experts," the report read, "recommends the control system described above for consideration by governments."[11]

This change in the purpose of the conference might not have been significant in itself if the American experts had not made certain other mistakes which persons more skilled in diplomacy might have avoided. In the first place, they failed to appreciate the propaganda value of the final report and communiqué. If they had had such an appreciation they might have seen to it that the weaknesses in the Geneva System were presented more forcefully. In particular, they should have insisted more strongly than they did that the final report of the experts indicate that the Geneva System had a *threshold*

[9] *ibid.* (Italics mine.)
[10] U.S. Congress, Joint Committee on Atomic Energy, *Hearings on Technical Aspects* . . . . , *op.cit.*, p. 19.
[11] *ibid.*, p. 526.

of detection-identification capabilities *below which it was relatively ineffective*. Instead, in the final communiqué and especially in the report, the major impression communicated to the world-wide audience was that the experts had disposed of the technical problems of a test-ban once and for all.

Secondly, the American experts concurred with conclusions which, although technically correct, ought to have been more emphatically qualified. It appears that instead, phrasing was chosen which blurred points of disagreement on the capabilities, limitations, and characteristics of the control system and thus permitted East-West agreement. As a consequence of this failure to specify and to define the terms in the report precisely, the scientists opened the door for subsequent disagreements between East and West in the political negotiations over the meaning of important statements in the report. For example, a major point of conflict between East and West since 1958 has been the phrase in the report that "there would always be a possibility of detection" of a covert underground test. This crucial phrase in the technical agreement can be, and has been, interpreted to mean anything from nearly zero per cent to 100 per cent probability of catching a violator.

Thirdly, the failure of the experts' report to emphasize the limitations of the system permitted and even fostered undue optimism; actually the impression created at a number of points in the report was in marked contrast with the technical reality as seen by the American scientists themselves. The most striking example of this was the critical problem of on-site inspection of underground nuclear explosions. For, in spite of a highly pessimistic analysis by one of the American experts at the conference that under certain circumstances "proof of a violation by technical means . . . would be lacking,"[12] the final report of the conference stated only that "a violator . . . could not be guaranteed against exposure, particularly if . . .

---

[12] Conference of Experts . . . . , *op.cit., Verbatim Transcript of Sixteenth Meeting,* July 21, 1958, pp. 81–85.

[there is an] inspection at the site of the suspected explosion."[13] Apparently, the American experts assumed in agreeing to this that non-technical means such as intelligence activities would supplement the technical methods.

Lastly, the American experts may be criticized as diplomatic negotiators. Inexperienced and apparently poorly advised, they were outmaneuvered by their Russian counterparts and, unwittingly, they made some rather unfortunate political concessions which negotiators with diplomatic experience undoubtedly would not have made. These concessions, however, must have seemed relatively trivial to the American scientists when compared to the overall technical agreement which had been achieved.

The first substantive difference (between the Western and the Soviet delegations) which was compromised to the disadvantage of the control system was that which concerned the use of aircraft by the control organization. The American scientists initially took the position that, as regular aerial collection of radioactive debris around the world was far superior to the use of ground filters, aerial survey should be an integral part of the detection system in addition to collection through ground filters. The Soviet scientists objected on the grounds that the use of aircraft was "unscientific" and too difficult to carry out. They proposed instead that total reliance be placed upon ground filters.

The compromise agreed upon was to restrict regular aircraft flights to flights over the world's oceans and to provide that, under certain conditions, flights could also be used to investigate suspicious events. However, the "certain" conditions under which such flights could be made were left indefinite except that the inspected country would lay out the flight plan. A crucial point not even discussed was whether or not the inspected country could insist upon other necessary conditions prior to the dispatch of inspecting aircraft over its territory.

---

[13] U.S. Congress, Joint Committee on Atomic Energy, *Hearings on Technical Aspects* . . . . , *op.cit.*, p. 526.

However, one may argue that these limitations were not serious in that the primary method for air bursts is acoustic detection.

The subject of the use of aircraft by the control commission furnishes yet another example, and a striking one at that, of the manner in which the Russians maneuvered for political advantage in the "technical" talks. The following is from the official transcript:

> *Mr. Fedorov (interpretation from Russian):* Then, Dr. Fisk, I should like to suggest modification of the text which you have proposed. I would propose that after the words "open seas," which occur in two places, we add, in parentheses, the word "oceans." Why do I suggest this? Because the flights of aircraft for meteorological purposes—and the use of such flights for the collection of samples of radioactive products—are being carried out, and probably will continue to be carried out, specifically over oceans and not over small seas. Legally, everything may be considered to be an open sea—but, prac·tically speaking, such flights will take place over oceans.
> *Mr. Fisk:* In the interests of saving words, why do we not simply replace "open seas" by "oceans?"
> *Mr. Tsarapkin:*[14] "Over the oceans"—plural?
> *Mr. Fisk:* Yes, plural—"oceans."
> *Mr. Fedorov (interpretation from Russian):* Then, Dr. Fisk, may we consider that these conclusions have been accepted and approved by our conference? Very good.[15]

Obviously Fedorov, a scientist, had had coaching from an expert in international law. Similarly it would have been wise for the head of the American delegation to have inquired into the legal difference between "high seas" and "oceans"— especially as understood by the Russians—before agreeing to

[14] For Tsarapkin's identification see the conclusion of this chapter. The other Russian, Fedorov, was head of the Russian delegation.
[15] Conference of Experts . . . . , *op.cit.,* *Verbatim Transcript of Eighteenth Meeting,* July 23, 1958, pp. 6–11.

the Soviet suggestion. A change in legal meaning, and not the desire to save words, was certainly the purpose of the Russian request. This seemingly innocent exchange of terminology might well have meant that under the Geneva System no regular aircraft flights to collect radioactive debris would have been possible near the Soviet Union because the USSR is bounded solely by "seas" and is untouched by the world's "oceans" except at the lower extremity of the Kamchatka Peninsula.[16]

A second extremely important political problem which the American scientific delegation permitted the conference to leave unresolved was who or what would decide whether or not to initiate an on-site inspection of a suspicious event. The significance of this political issue for the technical deliberations was indicated by Fisk himself who told the conference, "the capabilities of the system . . . depend, importantly, on the provision of inspection, and on *how* this provision is carried out."[17] This issue of inspection would dominate all subsequent political negotiations.

The position of the United States on this issue was, and has been throughout the subsequent political negotiations, that scientific criteria should be established by which non-inspectable events could be separated out, i.e., by which events could be identified as earthquakes. All other events would then be eligible for on-site inspection. The Soviet scientists, on the other hand, insisted that the decision to initiate each on-site inspection should rest solely with the control commission upon which the nuclear powers would sit. Furthermore they have maintained that, while the commission could use technical criteria in making such decisions, these criteria should be such that they would identify particular events

[16] Russian sensitivity with respect to foreign trespassing on its peripheral "small seas" and its claims to sovereignty over adjacent seas are discussed in William N. Harben, "Soviet Attitudes and Practices Concerning Maritime Waters—A Recent Historical Survey," *The JAG Journal*, Vol. 15, No. 8, October–November 1961, pp. 149–54.
[17] Conference of Experts . . . . , *op.cit.*, *Verbatim Transcript of Twenty-Eighth Meeting*, August 7, 1958, p. 81. (Italics mine.)

as "suspicious." Then the control commission would determine whether or not a particular suspicious event would be inspected.

Thus, the actual issue dividing Western and Eastern scientists was whether or not the right of on-site inspection would be a guaranteed provision of the Geneva System. This is to say, would the Western powers have the *right* of inspection of Soviet territory? Unfortunately, the experts did not resolve this crucial political-technical issue. Instead, the ambiguous language in which the provision for on-site inspection was worded permitted East and West to draw opposed conclusions from it. In one place the final report read, "for those cases which remain unidentified inspection of the region will be necessary" and, at another point in the report, "[the control commission] *can send* an inspection group . . . in order to determine whether a nuclear explosion had taken place or not."[18]

For the Western experts this language was regarded as a major victory over Russian intransigence on the inspection issue; it was believed that, for the first time in the postwar disarmament negotiations, the Russians had agreed to meaningful inspection. As subsequent events were very soon to reveal, however, the Russians intended that the Great Powers on the control commission, as in the Security Council of the United Nations, would exercise a veto over all substantive matters including the initiation of an on-site inspection. They have yet to veer from this position throughout the test ban negotiations.

It would appear that prudent negotiators, considering the importance of the issue itself and in the light of previous experience with the Baruch Plan and other disarmament negotiations, ought to have recognized the need to establish whether or not the West would have an unequivocal *right* to inspect events it regarded as suspicious. If the experts had

---

[18] U.S. Congress, Joint Committee on Atomic Energy, *Hearings on Technical Aspects* . . . . , *op.cit.*, pp. 488, 524. (Italics mine.)

succeeded in clarifying this point at the time, the West might have been able to determine early in the negotiations whether the USSR had *really* accepted the principle of international inspection.

A further error in the diplomacy of the American experts was their failure to assume that any nation which might seek to violate the control system by covert means would do so with resourcefulness and determination. As a consequence, the scientists did not really evaluate the control system in terms of its capability of preventing evasion by a skillful violator. In the words of Harold Brown the "Conference of Experts report can logically be criticized for not taking adequate account of the self-evident fact that a violator of a treaty or a moratorium on nuclear weapons tests must be expected to take full advantage of whatever methods he can find to reduce the probability of detection. It is therefore rather misleading to consider only the probability of detection of an unconcealed shot."[19]

The Reaction of American Scientists to the Conference of Experts

The reactions of American scientists to the Conference of Experts and, of course, to President Eisenhower's decision to seek a nuclear test ban are significant factors for the subsequent history of the intra-scientific struggle over a nuclear test ban, for these events reinforced the split within the American scientific community over the wisdom of a nuclear test ban. On one side are ranged the scientists in the control and finite containment schools who favor a nuclear test ban; on the other are the infinite containment scientists who are opposed.

THE REACTION OF THE PRO-BAN SCIENTISTS

To the scientists in the control and finite containment schools, the successful conclusion of the Geneva Conference

[19] *ibid.*, p. 35.

of Experts demonstrated the willingness of the Russians to talk seriously about inspected disarmament for the first time since World War II; these men believed that the consequent opportunity to take some step toward disarming the world should be exploited to the full. Furthermore, these scientists viewed the conference as the first in a long series of technical talks which would solve, one at a time, the many problems of disarmament. Already there had been scheduled a Conference on Methods to Prevent Surprise Attack to commence November 10, 1958. Hopefully, this conference would be followed by technical conferences between East and West on nuclear production controls, conventional arms reductions, and prevention of biological warfare.

From the perspective of these scientists, the Conference of Experts had proved their case that the scientists of the world could succeed where the statesmen had failed. The international language of mathematics, the spirit of scientific co-operation, and the objectivity of searchers after truth had been able to replace propaganda, maneuver for advantage, and parochial viewpoints. Indeed, the scientists of East and West had shown the nations the way in which to take the "first step" toward disarmament through their successful substitution of the criterion of technical feasibility for that of "national interest."

A fitting summary of this attitude of many pro-ban scientists was provided by an editorial in the *Bulletin of the Atomic Scientists:* "The success of the Geneva conference of scientists . . . has confirmed the belief of scientists that once an international problem has been formulated in scientifically significant terms, scientists from all countries, despite their different political or ideological backgrounds, will be able to find a common language and arrive at an agreed solution. . . . This new approach to the fundamental problems of the arms race and world security, using the criterion of what is technically the most feasible approach to a common aim, instead of what will satisfy the national interests, may be a more radical innovation than the political leadership of the major nations is now willing to contemplate, despite the disastrous experience of traditional diplomacy in the last ten years."[20]

While most scientists who favored a nuclear test ban would probably not have gone as far as the statement above, there is no doubt that these scientists believed the moment was propitious for progress in arms control and, eventually, in disarmament. Many believed that not only had a major step been taken along the path toward nuclear disarmament, but that the conference and its results would make the nations appreciate the value of the scientific method and of the international scientific community as major instruments for achieving lasting peace. Shortly after the conclusion of the Conference of Experts, the Council of the American Association for the Advancement of Science expressed this sentiment: "We believe that these negotiations [to implement the report of the Conference of Experts] represent a bright hope for the translation of scientific knowledge into effective public policy on a question which—literally—involves the survival of civilization."[21]

The pro-ban scientists did realize, however, that the Geneva System was far from foolproof. Nevertheless, they believed that the system's acknowledged weaknesses would be reduced through practical experience and the advance of knowledge. As the final communiqué of the experts put it, "the effectiveness of the methods considerably increases in the course of time, with improvement of measuring techniques and with study of the characteristics of natural phenomena which cause interference when explosions are detected."[22]

The pro-ban scientists found another cause for optimism about future developments in the excellent rapport between

<hr>

[20] Eugene Rabinowitch, "Nuclear Bomb Tests," *Bull. Atom. Sci.,* Vol. 14, No. 8, October 1958, p. 287.

[21] "Resolution on Control of Nuclear Weapons Tests," *Science,* Vol. 129, No. 3342, January 16, 1959, p. 137.

[22] Conference of Experts . . . . , *op.cit., Verbatim Transcript of Thirtieth Meeting,* August 21, 1958, pp. 4–6.

Western and Eastern scientists which existed at Geneva. The reuniting of ties with Russian scientists which had begun at the 1955 Atoms for Peace Conference appeared to have borne fruit. In the eyes of these men it appeared that the international language and community of science had made possible the achievement of agreement on a subject long in dispute among the diplomats: an inspected arms control agreement. On the basis of the Conference of Experts the pro-ban scientists were convinced that Soviet scientists would cooperate with them to make the Geneva System work and that the Great Powers had begun to build that mutual trust whose lack was regarded by many scientists to be the major impediment to the international control of atomic energy.

THE REACTION OF THE ANTI-BAN SCIENTISTS

In decided contrast to this optimistic view concerning the Conference of Experts was the reaction of the infinite containment scientists. These scientists thought that the American experts at Geneva, either through their naïveté or their strong desire for a nuclear test ban, had permitted the Russian scientists to maneuver the United States into a *cul de sac* where it had no choice but to agree to a test ban treaty or to suffer a terrible propaganda defeat.[23]

These scientists believed that the American delegation, or at least some of its leaders, had permitted their desire for a nuclear test ban to overpower their commitment to scientific objectivity. As a result they thought that the American experts had agreed to a control system of extremely questionable capabilities and had misled the world with respect to the technical feasibility of a nuclear test ban. For example, Freeman Dyson, although not a member of the infinite containment school, has stated this criticism in a moderate form: "The experts at Geneva in 1958," Dyson wrote in 1960, "did not

[23] For a statement of this argument by a non-scientist see Thomas Murray, "East and West Face the Atom," *The New Leader,* June 15, 1959, pp. 10–14.

know to what extent artificial concealment of nuclear explosions was technically possible. Probably for this reason they decided to say nothing about it. *Scientific objectivity would, however, have required them to report that the concealment problem had not been explored.*"[24]

The scientists in the infinite containment school saw no indication in the behavior of the Russian scientists at Geneva of a new era of East-West relations wherein scientists would play a major role in an eventual rapprochement. Instead these scientists believed that the purpose of the Soviet Union at Geneva had been to create an international atmosphere which would at least inhibit further American nuclear testing and would perhaps deter American employment of tactical nuclear weapons. At the same time they feared that the Russian possession of territorial secrecy would permit them to continue nuclear weapons testing and, one day, to achieve decisive nuclear weapons supremacy over the West.

Furthermore, the scientists who opposed a nuclear test ban believed that the advance of knowledge would soon prove the inadequacies of the Geneva System. While they tended at first to keep these thoughts to themselves, they set to work to prove their case that disarmament has become an impossibility without world government. In particular, they hoped to show that, both in principle and in fact, "in the contest between the bootlegger and the police, the bootlegger had a great advantage."

Conclusion

In the mistaken belief that one can separate the technical and political aspects of national policy, American political leadership in the summer of 1958 assigned to a group of inexperienced private citizens the task of negotiating the first part of what might have been an extremely important arms control

[24] Freeman Dyson, "The Future Development of Nuclear Weapons," *Foreign Affairs,* Vol. 38, No. 3, April 1960, p. 461. Actually, Dyson's criticism is incorrect. The experts discussed and reported on the problem of evasion. The mistake of the American delegation was the political one of not emphasizing this problem sufficiently in their report. (Italics mine.)

agreement. As should have been expected, the American scientists at the Geneva Conference of Experts, lacking sufficient political guidance, fell into a number of regrettable errors. Yet, under the circumstances, it is surprising that there were no greater errors committed and that the scientist-diplomats did as well as they did.

The American experts at Geneva had been placed in a position where, of necessity, they had to use political judgment. In some cases, the scientists' political judgments were without doubt influenced by their desire to see a permanent ban on weapons testing; in other cases, they were influenced by the belief that they had a clear mandate from the President to reach a technical agreement if such were possible. Whichever the case and however much one questions the political judgment of the scientists, it would be quite wrong to impugn their integrity or to charge that they harmed national security.

Although the Western experts made a number of errors as diplomatic negotiators, none of these errors were necessarily serious in themselves. They only became serious when they were compounded by the numerous errors committed subsequently by American political leadership. In particular, the mistakes made by the scientists at Geneva would have been of relatively little consequence if American political leadership had not made the error of imposing upon itself a moratorium on weapons testing at the commencement of the political negotiations on October 31, 1958. With this announcement, however, the United States, believing the scientists had made a breakthrough on the inspection issue, placed itself in a difficult political position. The Russians then had what they wanted—an unpoliced moratorium on testing; moreover, through subsequent diplomatic maneuvers, they would succeed in prolonging it. The United States was soon to discover that, due to the ambiguous language and unresolved details in the experts' report, it did not have a Russian commitment, even in principle, to international inspection. If the United States had not chosen to suspend its tests in the fall of 1958 it could not only have continued the negotiations at no cost to its own nuclear weapons program, but the Russians would undoubtedly have had a greater incentive to negotiate in earnest.

The errors of the American experts in the negotiations can perhaps be laid to the fact that they were blinded by their apparent success as well as to the effect of inexperience, poor guidance, and a great desire to succeed. The writer speculates that at the time of the negotiations these scientists assumed that the Russians were primarily, if not solely, interested in large yield nuclear weapons. As the American scientists believed that in the negotiations the Russians had agreed to a control system which would have effectively policed tests of such weapons, the Western experts believed they had succeeded in their mission. Consequently they were negligent with respect to "details" such as the legal meaning of various terms and other fatal weaknesses in the final report. Furthermore, since the Administration did not know what it wanted from the negotiations, it could not properly guide the scientist-negotiators. The responsibility, therefore, for the errors committed at Geneva lies not with the scientists alone, but also with the Administration for its failure to guide the scientist-negotiators.

As for the American experts themselves, it was only much later that they began to appreciate the fact that in diplomatic negotiations, it is impossible to draw logical conclusions from technical facts in isolation from political considerations. Many months later, James Fisk himself was to make this observation: "The technical content of arms control negotiations is likely to be very high but experience in the nuclear test and surprise attack negotiations has shown that technical and political arguments cannot be separated completely or for long. For example, in the nuclear test business the questions of 'threshold' and probability of detection and identification have both technical and political matters deeply intermixed; if the technical people talk of a 1 kt [kiloton], 5 kt or 20 kt 'threshold' there are important political overtones; whether they talk of 10% probability or 50% or 90% is largely a political matter—or is it technical?" Then, Fisk concluded "I believe the Soviets recognized this interplay [of the technical and the political] from the beginning."[25]

In particular, there was one member of the Russian delegation who must have appreciated the interplay of the technical and the political far more than did the amateur American scientist-diplomats or, apparently, even American political leadership. There can be little doubt that the diplomatic skill of this man lay in large measure behind the maneuvers in the negotiations of the Russian scientists. It was he, for example—as the transcript of the conference reveals—who advised Fedorov to have "oceans" exchanged for "high seas" in the final report of the experts. For this reason, let us have him introduced by Charles Thayer, a former American professional diplomat who himself has experienced the deft negotiating skill of this Russian professional, and, therefore, might well appreciate what happened to the American scientific delegation at Geneva: "In 1958 a conference was convened in Geneva to investigate methods of detecting nuclear explosions. It was billed as a purely scientific exploration from which all political considerations were to be resolutely excluded. When the Soviet delegation stepped from its plane it was headed by a shaggy-haired little man with an unprepossessing manner and a crooked smile. You could have searched in vain for his name in every register of Soviet scientific institutions. No American scientist had ever read one of his papers or heard him address a scientific gathering. But he was well known to many American diplomats as one of the Kremlin's toughest negotiators . . . with the name of Simyon [Semyon] Tsarapkin."[26]

[25] U.S. Senate, *Strengthening the Government for Arms Control,* Senate Document No. 123, 86th Congress, 2nd Session, 1960, p. 7.

[26] Charles Thayer, *Diplomat,* Harpers, 1959, p. 106. Tsarapkin's experience in disarmament negotiations goes back to the Baruch Plan negotiations of 1946. He was not the head of the Russian delegation to the Conference of Experts, but he, a non-scientist, was a member of it; while there were Western political observers at the conference, they were not actually members of the Western delegations, and they apparently played a minor role in the negotiations.

"Old Scratchy," as Tsarapkin came to be called, would become well known to the American public within a few months when he emerged from the background of the technical talks to become the wily, resourceful head of the Soviet delegation to the political negotiations on a test ban treaty. But by then it was too late; Tsarapkin had only to maintain in the "political" talks the position on inspection he and his colleagues had already established in the "technical" ones.

CHAPTER EIGHT
THE "FIRST STEP" FAILS[1]

PRESIDENT EISENHOWER'S decision, based on the "successful" Geneva Conference of Experts, to negotiate a nuclear test ban treaty made critical the contradiction in American nuclear policy. On the one hand, American armament policy was based firmly on the doctrine of massive retaliation while, at the same time, the United States was rapidly developing its limited nuclear war capabilities. On the other hand, the United States had now embarked upon a new arms control policy whose immediate effect, if successful, would be to arrest the further development of tactical atomic weapons and whose ultimate consequence could well be to outlaw the nuclear weapons upon which the United States had based its defense. The United States could not long subscribe both to the continued development of a capability for limited nuclear war as desired by the infinite containment scientists and to a total ban on nuclear weapons testing as envisioned by other scientists. In the perspective of each set of scientists, the future of mankind rested upon the outcome of the debate, and the struggle among the American scientific proponents and opponents of a nuclear test ban was to become a significant factor in the larger attempts of both the Eisenhower and the Kennedy Administrations to establish such a treaty.

The attempt to implement the Geneva System, however, was to encounter a number of difficulties. The anti-ban scientists were to make a strong case against the technical feasibility of the Geneva System, and new scientific data from

[1] On the attempt to implement the Geneva System, detailed histories which differ in interpretation from the present writer's account are Bernhard G. Bechhoefer, *Postwar Negotiations for Arms Control*, The Brookings Institution, 1961, and Department of State, *Geneva Conference on the Discontinuance of Nuclear Weapons Test—History and Analysis of Negotiations*, October, 1961.

nuclear explosions conducted after the conclusion of the Conference of Experts would reveal that the actual capabilities of the Geneva System were far inferior to those assumed by the experts. As a consequence, the problem of how to improve the system in order just to bring it to its initially assumed capabilities would be raised. Even more significant in its effect upon the feasibility of the Geneva System would be the behavior of the Soviet Union. Contrary to the expectations of the pro-ban scientists, the Soviet Union and its scientists were to refuse to acknowledge that the new data meant a decrease in the actual capabilities of the Geneva System. The Russians would neither concede the necessity of improving the system up to and beyond its originally assumed capabilities nor would they agree to the implementation of the Geneva System.

Nevertheless, the basic faith of the pro-ban scientists that the "logic of the facts" of nuclear disarmament would eventually be accepted and implemented by the nations has continued to remain firm. This is so even though they have come to realize that the Russians were more intractable on the subject of inspection than these scientists had believed them to be immediately following the Geneva Conference of Experts. Moreover, although the position of the anti-ban scientists would be largely confirmed by subsequent events, the need for some measure of arms control would be increasingly recognized.

The Attempt to Implement the Geneva System:
Phase I

The opposition of the infinite containment scientists to the test ban was soon strengthened by both technical and political events. In the former category were a new series of American nuclear explosions which exposed serious weaknesses in the technical assumptions underlying the Geneva System. In the latter category was the refusal of the Soviet Union to agree to international control as part of a treaty permanently banning further nuclear tests. Together these factors were to make impossible the successful implementation of the report of the Conference of Experts.

DETERIORATION OF GENEVA SYSTEM CAPABILITIES

Project Argus and the Johnson Island shots. At the time that the Conference of Experts was concluding its sessions, the United States was conducting two series of nuclear explosions high above the Pacific Ocean which were to have serious implications for the conclusions of the experts. The first set of tests, known as the Johnson Island shots, was conducted in July and August of 1958. There were two shots of megaton yield which were exploded at less than 100 kilometers. The second series, entitled Project Argus, consisted of three 1 to 2 kiloton explosions; they were exploded at 500 kilometers. Thus, all five explosions were above the upper reaches of the proposed Geneva System techniques, yet the fact that these explosions had taken place was not revealed until March 1959 when *The New York Times* carried the story.

The success of the Johnson Island and Argus shots centered American and world attention on the fact that the Geneva System would not detect nuclear explosions at very high altitudes. Furthermore, the American tests made it obvious that it was no longer appropriate to "brush aside" this type of testing as unimportant, as the Conference had done. High altitude testing was not only feasible, but it had now been proved that it could be accomplished at a reasonable cost.

The Hardtack series.[2] Another series of nuclear explosions, underground this time, was held shortly after the conclusion of the Conference of Experts but prior to the convening of

---

[2] The underground shots listed below were part of the Hardtack series: Evans (55 tons), Tamalpais (72 tons), Neptune (90 tons), Logan (5 kilotons), and Blanca (19 kilotons). See G. W. Johnson, G. H. Higgins, and C. E. Violet, "Underground Nuclear Detonations," *Journal of Geophysical Research,* Vol. 64, No. 10, October 1959, p. 1461.

the political conference at Geneva. Then, while the political negotiators met in the Conference on the Discontinuance of Nuclear Weapons Tests throughout the period from October 31 to December 19, American technical experts studied the new data provided by the Hardtack series. By January 1, 1959 it was evident from this new data that the capabilities of the Geneva System were much lower than the experts had judged them to be. On January 5, 1959 the United States released a statement which summarized the significance of the new data: "[Recent analysis] including data new since last summer, has shown that this method of distinguishing earthquakes from explosions is less effective than had been estimated by the Geneva Conference of Experts. These analyses and new data also indicate that the seismic signals produced by explosions are smaller than had been anticipated and that there are consequently about twice as many natural earthquakes equivalent to an underground explosion of a given yield as had been estimated by the Geneva Conference of Experts.

"These two factors mean that there will be a substantial increase in the number of earthquakes that cannot be distinguished from underground nuclear explosions by seismic means alone. For example, the total number of unidentified seismic events with energy equivalents larger than 5 kilotons may be increased ten times or more over the number previously estimated for the system recommended by the Geneva Conference of Experts."[3]

Originally it had been estimated that the Geneva System would be capable of identifying as earthquakes 90 per cent of the events which registered an equivalent of a 5 kiloton explosion or above. The new data indicated that the Geneva System "would have about the same capability (in terms of numbers of unidentified events) for seismic events above 20 kilotons equivalent as was originally estimated . . . for seismic events above 5 kilotons." In summary, this meant that within

[3] Department of State Press Release, January 5, 1959.

the Soviet Union many seismic events less than 20 kilotons equivalent as well as about seventy events per year above 20 kilotons would be unidentifiable as earthquakes without on-site inspection.

THE RESPONSE OF THE UNITED STATES GOVERNMENT AND THE PRO-BAN SCIENTISTS

The response of the United States Government. The initial response of the Eisenhower Administration to the revelations of the Hardtack data was to call upon the Soviet Union to agree to a technical reassessment of the Geneva System. Specifically, the United States desired a reconvening of the Conference of Experts in terms of the new American data. Then the experts could recommend what measures should be taken to return the Geneva System to its initial capabilities.

The immediate Russian reaction to this proposal was the charge that the new data were fabricated as part of a plan to discredit the Geneva System. The American purpose, the Russians held, was to prevent the consummation of a treaty banning further nuclear tests. For this reason, they argued, they would have nothing to do with the new American data. And although they did finally agree in late 1959 to review the Hardtack data, they have yet to admit its relevance to the capabilities of the Geneva System.

The second American response to the discouraging setbacks suffered in the test ban negotiations was to make a further concession to the Soviet Union. On January 19, 1959 President Eisenhower abandoned one of the qualifications made in his August 22 announcement calling for negotiations to implement a nuclear test ban. No longer would American adherence to a nuclear testing moratorium be contingent upon "satisfactory progress . . . being made . . . [on the] substantial arms control measures such as the U.S. has long sought." The purpose of this move, according to a "high U.S. official," was to demonstrate the West's "sincere determination for agreement" and, in rather blunt language, to place the USSR in a "put-up or shut-up position."[4]

The response of the pro-ban scientists. Although the technical and political developments in late 1958 were discouraging, they did not daunt the faith of the pro-ban scientists in the eventual achievement of a nuclear test ban. In fact, although the new data showed that the conclusions of the experts had been overly optimistic, the findings seemed to provide at least a solid foundation of knowledge upon which the system could be refashioned and improved with the assistance of Soviet scientists. After the release of the Hardtack data, Bethe stated with regard to underground testing, "Now I feel there is enough evidence [to evaluate fully the problem of detection and identification] in particular since one of the [October] shots was quite high—about 20 kilotons." He added, "A shot as big as that can be observed all over the map. Stations were alerted beforehand. Extra seismographs were put up. We have very much evidence.

"We would like to have other tests. But we are not sure whether we should blow up this desire too much. It might then be difficult to get the Russians to react in a constructive way."[5]

In particular, Bethe's statement revealed the strong disposition of many pro-ban scientists to believe that the Soviet Union, commencing with the Conference of Experts, had entered a rare mood of rationality; it was the feeling of those holding this view that the West should take advantage of this possibly temporary aberration in Soviet behavior to erect a stable arms limitation system but that care would be necessary to avoid moves which would cause the Russians to revert to their former irrational position.

STALEMATE IN THE POLITICAL NEGOTIATIONS

With respect to the political talks on a test ban treaty, there was little room for Bethe's apparent optimism. Although the

---

[4] *The New York Times,* January 20, 1959, p. 1.
[5] Quoted in the *Washington Star,* January 11, 1959.

negotiations continued from October 31 through December 19, 1958, the negotiators reached agreement on only four rather minor treaty articles. All major issues were in deadlock with little hope for compromise. In particular, three important matters were distressing to the United States: (1) Russian insistence upon a veto over all substantive questions to be decided by the control commission; (2) Russian imposed restrictions on international on-site inspection of suspicious events; and (3) Russian insistence that control posts be staffed by citizens of the country in which the posts were located—with only one or two outside "observers" allowed.

In essence, the Russian position in the fall of 1958 was that the control system meant "self-inspection plus veto." The Soviet delegation, led by Tsarapkin, not only insisted that the Soviet Union would have a veto power over the initiation of on-site inspection within Russia but that the inspectors themselves and the staffs of the control posts within Russia should be Russian nationals. The Western powers would be permitted only to attach "observers" to inspection teams and control post staffs in the USSR. These basic demands of the USSR were reinforced by other qualifications which would in essence have negated the inspection system.

Throughout these political negotiations the Russians enjoyed the advantage of being able to make references to the report of the experts in support of their political demands. Astonishing interpretations of various sections of the report were provided by the Soviet delegates. In addition, there was a change in Russian demeanor. Whereas during the technical talks the Russian delegates had been informal and obliging, they were now formal, legalistic, and uncompromising in their interpretation.

Not only had the test ban negotiations run into serious difficulties, but the East-West technical conference on the prevention of surprise attack which had commenced November 10, 1958 was also stalemated. Then, after five weeks of further negotiations, these talks collapsed. While the inadequate preparation of the American delegation was a significant factor in this development, the attitude of the Soviet Union was the major cause for concern. The Russian delegates refused even to discuss the technical aspects of the subject unless the United States agreed to three conditions: (1) the abolition of nuclear weapons; (2) elimination of all foreign bases; and (3) reduction of all conventional arms. As the United States refused to discuss these non-technical points, the talks were recessed indefinitely and as this is written they continue to be in recess.

The Attempt to Implement the Geneva System:
Phase II

In response to the new data provided by the Hardtack series, Killian, on December 28, 1958, had appointed a Panel on Seismic Improvement to make recommendations on ways in which the Geneva System could be improved. The panel, usually referred to as the Berkner Committee after its chairman, Lloyd V. Berkner, was to address itself to three main problems: (1) whether presently available techniques could increase the efficiency of the proposed 180-station detection net; (2) whether research in seismology could lead to further improvements; and (3) whether it were possible to muffle underground explosions, increasing the difficulty of distinguishing them from earthquakes by seismic readings.

After an extensive study, the Panel on Seismic Improvement presented its findings on June 12, 1959, to the Conference on the Discontinuance of Nuclear Weapons Tests. The panel's first conclusion was that a proper use of available techniques could improve the system. One such technique, for example, would be to increase the number of seismometers at each control station from ten to one-hundred and thereby make it easier to distinguish seismic first motions. Another improvement could be achieved by the use of surface seismic waves to aid in the separation of explosions and earthquakes. The effect of this type of improvement would be to give the 180-station net the same detection and identification capability for earthquakes with an energy equivalent to 10 kilotons and over that the system had originally been estimated to have for events equivalent to 5 kilotons and over. However, the greatest boon for the improvement of the system, the panel reported, would be utilization of a supplementary system of unmanned or robot stations in addition to 180 manned seismic stations. A network of these robot stations spaced 170 kilometers apart in seismic areas would identify 98 per cent of the seismic events of 1 kiloton equivalent and above.

With respect to future improvement through research in seismology, the panel believed that several ways could be developed to improve the efficiency of the system. The use of seismometers in deep underground holes and the employment of computers to reconstruct original wave forms were suggested. It was also noted that a testing program utilizing nuclear and chemical explosives might produce other feasible improvements.

The panel had also been asked to study the problem of evasion since the possibilities of evasion would seriously affect any program for seismic improvement. The conclusion of the panel on this subject was that the use of evasion techniques could decrease by at least ten-fold that energy of an explosion which is converted into seismic waves. As a result, a 10 kiloton explosion could be made to register on a seismograph as a 1 kiloton event. Techniques for evasion could include certain ways of locating the explosion, designing the explosion chamber, and varying the medium in which the explosion takes place.

Throughout the political negotiations at Geneva, the Soviet Union continued to insist both that the Geneva System as drawn up by the experts was adequate to monitor a nuclear test ban and that the USSR ought to have the right of veto with respect to on-site inspection of Russian territory. In an effort to resolve the East-West impasse over the adequacy of the control system, President Eisenhower, on April 13, 1959, proposed, in a letter to Khrushchev, a treaty which would ban only *policeable* tests. Such a partial ban would prohibit testing in the atmosphere up to 50 kilometers and in the oceans. The President's plan further suggested that the ban could be extended to all environments as inspection techniques improved.

On April 23, the Russians, arguing that *all* tests had been shown by the experts' report to be policeable, rejected the President's proposal. Instead, the Soviet Union continued to press for a total ban on testing and argued that the American proposal was calculated to mislead the public into a belief that agreement on a ban had been reached when the United States, under its proposal, could actually continue to test underground and in space. In addition, the Soviet Union took up a suggestion first made by British Prime Minister Harold Macmillan during a visit to Moscow that the nuclear powers should negotiate an annual quota of on-site inspections. Significantly, however, the Soviet Union qualified this apparent concession with the provision that inspection of any particular event could be initiated only if that event exhibited certain positive characteristics which would lead to suspicion that it was an atomic explosion. Thus, although the Soviets appeared to drop their veto on inspection, it was retained in the guise of a limited inspection quota and the requirement of "objective instrument readings" to initiate an inspection.

In a letter to Khrushchev dated May 5, 1959, President Eisenhower expressed his willingness to explore Khrushchev's proposal for the establishment of an annual quota of inspections; however, the letter also sought further clarification of the new Soviet position and specified that the size of the quota of inspections would have to be in appropriate relationship to scientific facts and to detection capabilities. In addition, President Eisenhower proposed a technical conference to devise a control system for tests at high altitudes and in space (above 50 kilometers). Unfortunately, one effect of these technical discussions and further attempts to draw out the Russians on the meaning of their April 23 counterproposal was to obscure the President's proposal of April 13 and to push it into the background.

In response to President Eisenhower's request for technical talks on high altitude and space detection, Russian and Western delegates met in Geneva in the early summer of 1959. Despite Russian obstinacy on a number of points such as a refusal to accept radar as part of the control system, Technical Working Group I (TWGI), as the talks were called, achieved considerable success. In addition to recommending the establishment of a system of earth satellites and ground-based equipment to monitor space testing, the scientists in TWGI produced for the first time an agreed estimation of the difficulties of policing a ban on high altitude and space testing.

In effect, the report of the TWGI stated that only one detection technique, thermal X-ray, was applicable at distances beyond 200,000 miles[6] and that even this method's effectiveness could be significantly reduced by the use of shielding materials surrounding the test shot. Given sufficient incentive to incur the cost of shielding, rocketry, and instrumentation, a nation could test nuclear weapons with a yield as large as .5 megaton (500 kilotons) beyond 200,000 miles with little or no likelihood of detection even if a rather elaborate detection system were in existence. In other words, as the head of the United States delegation to TWGI, Wolfgang Panofsky, put it somewhat later, "From the purely technical point of view, it appears likely that, given arbitrarily high incentives on the part of the violator, it will always be possible to devise an essentially undetectable means of carrying out the violation."[7]

Despite the successful conclusion of TWGI, the Soviet

---

[6] This is less than the average distance (239,000 miles) from the earth to the moon.

[7] U.S. Congress, Joint Committee on Atomic Energy, *Hearings on the Technical Aspects of Detection and Inspection Controls of a Nuclear Weapons Test Ban,* 86th Congress, 2nd Session, 1960, p. 37.

Union continued to reject the American demands for a reassessment of the Geneva System capabilities. For the Russians the experts' report was an inviolable international agreement and as such the only permissible changes in the technical assessment and characteristics of the Geneva System were those already provided for in the report. In effect, this meant the Russians would acknowledge the validity of only that technical data which justified and supported the original assessment made by the experts. This interpretation limited the range of permissible technical discussions principally to matters of instrumentation and instrument readings.

The United States and Great Britain, on the other hand, insisted that the Geneva System itself be re-evaluated in the light of all new data. In their view, the experts' report was not "a formal agreement on technical facts" but a "technical working paper,"[8] and as such it ought to be revised as knowledge of the scientific facts developed. Sincerely or not, the Russians argued that the Western Powers were trying to renege on a properly negotiated international agreement.

It must be noted that in addition to these difficulties in the technical realm, the political side of the negotiations provided little basis for optimism. Although by June 1959 the number of articles agreed upon in the draft treaty had grown to seventeen, the major issues under contention remained relatively unchanged. The Russian position on the veto, on-site inspection, and staffing of the control commission had not changed sufficiently to give much encouragement to the Western delegations. And soon these political and technical obstacles were to be reinforced by yet a third difficulty, the criticism of the Geneva System by the infinite containment scientists.

The Attempt to Implement the Geneva System: Phase III

Since his defeat on the question of the advisability of negotiating a nuclear test ban, Edward Teller had made few

[8] *ibid.*, p. 57.

statements on the matter. Nevertheless, to his colleagues in Berkeley and undoubtedly to sympathetic persons in the Eisenhower Administration, he expressed his deep concern over the weaknesses of the Geneva System. Teller believed that, in addition to the original limitations of the Geneva System and those revealed by the Hardtack, Argus, and Johnson Island tests, the continued advance of knowledge would cause still further degeneration of the validity of the experts' conclusions.

In particular, Teller was concerned over the possibility that a nation could evade detection through the use of methods to decrease that energy of an underground explosion which was converted into a seismic wave and thereby reduce the magnitude of the explosion's positive first motion. If such an evasion technique were sufficiently effective, the recording of an explosion could be masked by the background noise in seismographs or the explosion would be registered as one of many score of small yield disturbances with positive first motions.

This possible evasion technique, known as decoupling, had been briefly discussed and dismissed by Bethe at the Conference of Experts; in his presentation to the conference, he emphasized "that we severely lack good experiments" on decoupling, and immediately added that even his own view of it was "very crude theory only, and that one should not rely on it too much."[9] The experts did not believe decoupling was a serious enough problem to warrant much discussion or mention in their report. Instead, it will be remembered, they merely indicated that evasion methods did exist while noting that such measures were not effective enough to guarantee that a violator could avoid detection.

[9] Conference of Experts to Study the Possibility of Detecting Violations of a Possible Agreement on Suspension of Nuclear Tests, *Verbatim Transcript of Fourteenth Meeting,* July 17, 1958, pp. 82–85. The importance of the decoupling theory to the experts may be judged by the fact that Bethe was the only one to discuss it; his discussion occupied only one short paragraph of the transcript; and he discussed only one of a number of possible ways to decouple explosions.

Taking exception with the experts and with Bethe in particular, Edward Teller believed that the possibility of decoupling an explosion presented a formidable obstacle for an effective control system. He therefore suggested to a group of physicists at the RAND Corporation that they make a theoretical investigation of the problem of decoupling.

On March 30, 1959, the RAND physicists, under the leadership of Albert Latter, published their classified study. The summary of their report read: "It is shown theoretically that nuclear explosions can be effectively hidden in large underground cavities. An estimate of the effectiveness of the method indicates that a yield of more than 300 KT could be made to look seismically like a yield of 1 KT. Experiments with both chemical and nuclear explosions are needed to test the theory."[10]

The magnitude of this decoupling factor, if correct, meant, theoretically at least, a complete shattering of the Geneva System's capabilities. Given a cavity of sufficient size, explosions as large as 300 kilotons would undoubtedly go undetected. Such a revelation seemed to call for a radical revision of the system. Mere seismic improvements such as those suggested by the Panel on Seismic Improvement would be insufficient to counterbalance this possible evasion method; unless robot stations could be used extensively or a positive means to identify nuclear explosions could be discovered, it appeared that substantial improvement of the system would require a radical enlargement of the network of manned control posts.

The reaction of the advocates of the test ban to this revelation of the possibilities of decoupling as an evasion technique was one of intense hostility and dismay. They felt that it was irresponsible for a group of scientists to work to defeat the

[10] A. L. Latter, R. E. LeLevier, E. A. Martinelli, W. G. McMillan, *A Method of Concealing Underground Nuclear Explosions,* Santa Monica, Calif.: The RAND Corporation, March 30, 1959. Declassified October 20, 1959, p. 2. It should be noted that this group had studied another method for decoupling than the one discussed by Bethe at the Conference of Experts.

only promising negotiations on disarmament then in progress between East and West. The proponents of the test ban knew that added to the difficulties which had already appeared since the Conference of Experts, the vast opportunities for evasion opened up by the decoupling theory would provide valuable ammunition for all those persons both in and out of the Eisenhower Administration who were opposed to a treaty banning all nuclear tests.

Yet the decoupling theory withstood the test of scientific criticism. Bethe's independent calculations and those of other scientists convinced them that Latter's theory could not be discounted on theoretical grounds. Instead, therefore, many of these pro-ban scientists turned their attack upon the practical significance of Latter's theory. First within the Administration and then in public, following the declassification of Latter's study, these scientists sought to discredit the decoupling theory in various ways.

In general, the attacks on the "big hole" or decoupling theory, including some attacks purported to be engineering analyses, have taken the position that the cost, the time, and the effort necessary to construct a cavity sufficient to decouple explosions would be prohibitively high. In tone many of these "objective" studies have used the technique of ridicule. And in all the cases of which this writer has knowledge these critical analyses assumed, undoubtedly for polemical purposes, that the explosions to be decoupled would be at least 300 kilotons; the critics did not point out that *any* size explosion could be decoupled by a factor of 300 and that smaller yield explosions would require relatively smaller and less costly holes to decouple them.

Nevertheless, despite the refusal of many pro-ban scientists to admit the practicability of the decoupling theory, the advocates of the test ban along with the Eisenhower Administration did realize that the theory called for a serious reconsideration of the capabilities of the Geneva System. The decoupling theory in combination with the Hardtack data, which had not yet been discussed officially with Soviet scientists, certainly necessitated extensive additional discussions on the system's capabilities by the scientists representing East and West.

For these reasons the United States pushed more relentlessly for another technical working group to assess the actual capabilities and the improvements needed in the Geneva System on the basis of the Hardtack data and the decoupling theory. Finally, on November 25, 1959, approximately eleven months after the United States had begun to press the Soviet Union for technical talks on seismic improvements, the Soviet technical experts met with those from the West at Geneva. However, the terms of reference for Technical Working Group II (TWGII) were even more restrictive than had been those for Technical Working Group I. Whereas TWGI had been permitted by the Soviet Union to discuss and to recommend instrumentation for incorporation in the Geneva System, TWGII was to do no more than to "consider the question of the use of objective instrument readings" or criteria for use in the selection of suspicious events to be inspected. As in Technical Working Group I, this discussion was to proceed from "the discussions and the conclusions of the Geneva Conference of Experts."

Nevertheless, when the Soviet Union finally agreed after considerable resistance that "as part of their work" the experts could "consider all data and studies relevant to the detection and identification of seismic events" as well as "possible improvements of the techniques and instrumentation," the West was encouraged to believe that Technical Working Group II would be able to provide the needed evaluation and strengthening of the Geneva System. As the sessions of TWGII progressed, it became quite obvious, however, that the Soviet Union had not really revised its position on the conclusive nature of the Conference of Experts' report. As in the past, the report was considered by the

Russians to be inviolable, and the Russian scientists were adamant in their insistence that their American colleagues restrict themselves to a determination of "the objective instrument readings on the basis of which an on-site inspection can be initiated."[11]

Actually, the Russians were not as restrictive as the above statement indicates and they did tolerate a presentation by the American experts of the Hardtack data and the theory of decoupling. They refused, however, to discuss these findings properly and sought to discredit the new American data mainly through specious scientific arguments; when this technique proved impossible and embarrassing to them they shifted to the tactic of employing contrived technicalities. The Russians rejected the Hardtack data on the grounds that the seismographs used had not been those specified in the report of the Conference of Experts and that the seismographs had not been calibrated. Furthermore, they took the position that the Hardtack data were but the "opinions" of a few United States experts and, unlike the report of the Conference of Experts, not the conclusion of all parties to the Conference on the Discontinuance of Nuclear Weapons Tests.

The position of the American scientists was that the instrumentation used in the Hardtack series was the same that had been used in the Rainier test which had provided the original data for the experts with respect to underground tests. The data from the two test series were, therefore, comparable. Secondly, although the seismographs were not calibrated in either test series, this did not change the validity of the data. And if it did, it invalidated both the Rainier and the Hardtack data. Lastly, the data and conclusions from the Hardtack series were the "reasoned scientific conclusions" of the American delegation and ought to be treated as such by their Soviet colleagues.

The Soviet treatment of the decoupling theory defended

---

[11] It should be recalled that this matter was one left undecided by the Conference of Experts.

by Bethe and Latter was most condescending and even insulting. Chairman Fedorov of the Soviet delegation expressed to the Western scientists the attitude of the Soviet scientists with respect to the decoupling theory: "I will be very frank. I have absolutely nothing to do with your [decoupling theory] report. It is no longer of any interest to me."[12]

The official Russian position was far more diplomatic, although it meant the same thing. Essentially it is expressed in the view of another scientist of the Soviet delegation: "I have already put before you our view. I will repeat it: that, at the present time, the decoupling theory is speculative in nature and not confirmed by facts. Consequently, the Soviet delegation finds further discussion of it within this technical group premature. We have no proposal to make which would merit being given a place in the final report of the group."[13]

The lack of progress of Technical Working Group II toward a reassessment of the Geneva System on the basis of the Hardtack data and the decoupling theory was equalled by the failure of the scientist-delegates to reach agreement on the issue which the Soviet Union regarded as the central concern of TWGII. By the time of the convening of TWGII, it had become evident that the subject of "objective instrument readings" (on the basis of which the International Control Commission could initiate an on-site inspection) had become a crucial issue in the test ban negotiations. The decision of the USSR even to discuss such criteria seemed to be a shift from its prior position that it could exercise a veto in the control commission with respect to the inspection of Soviet territory. The new Soviet position, initially regarded as a "major concession" to the Western view, was that the experts should draw up a list of "objective instrument readings" to serve as the basis for initiating on-site inspection.

[12] Conference on the Discontinuance of Nuclear Weapons Tests, Technical Working Group II, *Verbatim Transcript of Twenty-First Meeting*, December 18, 1959, p. 61.

[13] *ibid.*, *Verbatim Transcript of Sixteenth Meeting*, December 14, 1959, p. 65.

Thus it appeared that the Soviets were proposing that objective criteria rather than a decision of the suspected nation would determine whether or not a suspicious event would be inspected.

This Soviet move appeared to go far to resolve an important stumbling block to the achievement of a nuclear test ban. It meant that the West would have a guarantee that it could inspect suspicious events in the USSR and that the Soviet Union for its part would not have to fear indiscriminate inspection by Western "espionage" agents. However, TWGII had not been in session long before it became obvious that East and West still had directly opposed notions on the subject of the initiation of on-site inspection.

The Western Powers believed that agreement should be reached upon objective criteria which would be used to eliminate from consideration by the control commission only the detected seismic events which could be positively identified as earthquakes. Then, all events which could not be definitely identified as earthquakes, i.e., events whose detected first motions were all positive, would be by definition suspicious and, therefore, eligible for on-site inspection. The Soviet scientists, on the other hand, desiring criteria which would keep the number of inspectable events as low as possible, maintained that only if a detected event had certain positive characteristics which virtually identified it as a nuclear explosion would it be eligible for on-site inspection.

The retort of the Western scientists to the Soviet position was that such positive criteria do not exist and that if they did exist there would then be no need to hold an on-site inspection. "I think," Dr. Carl Romney of the United States delegation told his fellow delegates, "the record is quite clear that there are no clear-cut definite criteria which [could] establish an underground event as a nuclear explosion except on the basis of an on-site inspection." Furthermore, the Americans pointed out, under the suggested Soviet criteria the explosions in the Hardtack series would not have been eligible for inspection. In fact very few, if any, seismic events would be inspectable.

Finally, after three weeks (November 25–December 18, 1959), the painful negotiations drew to a close. The time allotted to the scientists to resolve the differences between East and West had run out. Although a few minor issues had been solved, the major differences which had led to Technical Working Group II remained. The Soviet refusal to acknowledge the validity of the Hardtack data and the decoupling theory meant that basic disagreements continued between East and West on a number of critical technical points.

Prior to the breakdown of Technical Working Group II, the American scientists sought to re-establish with their Soviet colleagues the "objective spirit" which they believed had prevailed at the Conference of Experts. Wolfgang Panofsky spoke for his frustrated American colleagues when he appealed to their Russian counterparts for objectivity in the name of international science and world peace. "Nothing," Panofsky lectured his Russian counterparts, "is gained toward our common goal of achieving a controlled discontinuance of the testing of nuclear weapons by hiding facts known to us simply because the implications are not to our liking . . . . Unfortunately, there have been . . . a great many abuses at this meeting of the word 'objective' as applied to the examination of scientific results. It just should not occur that objective examination and a full exchange of views on seismic data should lead our Soviet colleagues to one conclusion while the conclusions of our own seismologists are different."[14]

Regrettably, Panofsky's remarks had little effect on his Russian scientific colleagues. In a separate annex to the final report of TWGII the Soviet scientists rejected out of hand the major contentions of the American scientists; they

---

[14] Conference on the Discontinuance of Nuclear Weapons Tests, Technical Working Group II, *Verbatim Transcript of Tenth Meeting,* December 7, 1959, pp. 4–6.

did not even consider it necessary to provide scientific refutation. In the last sentence of their statement they expressed their sentiments toward the position taken by the American scientists. "The Soviet experts," the annex concluded, "submit that here their United States colleagues are on the brink of absurdity."[15]

Thus, far from being the seekers after truth that the American scientists saw themselves to be, the Soviet scientists had proved themselves to be "politically motivated." Little did the American scientists realize, however, that the Russian scientists were actually no more political than they. The difference between the two sets of scientists was simply one of their respective political instructions. It was the political conflict between the scientific delegations which made agreement impossible on the relevant technical facts and on the implications of the facts for the Geneva System.

Nevertheless, the conviction that science is a force for peace is too entrenched in the thinking of many scientists to be reconsidered on the basis of the scientists' experience in these negotiations. Rather than concede that the Soviet scientists or any other scientist-negotiators will be guided by political instructions, many American scientists have viewed the behavior of the Russian scientists in these test ban negotiations as an aberration. In time, it is felt, their true scientific nature will win out over the narrow confines of Soviet ideology.

It is possible to be skeptical (as is the present study) of the notion widely held among scientists that the international community of science is somehow a force for peace; such skepticism does not deny the important role of scientists in the international sphere. Actually scientists will have an increasingly large part to play in international negotiations, especially if East and West embark upon serious arms control negotiations. The point is rather that neither the scientists nor

---

[15] U.S. Congress, Joint Committee on Atomic Energy, *Hearings on the Technical Aspects* . . . . , *op.cit.*, p. 615.

the responsible Western political officials should forget that such scientific discussions are equally as political in nature as they are technical. If the Conference of Experts, Technical Working Group I, and Technical Working Group II have proven anything, it is that the scientists can succeed only to the degree that the political situation permits.

Unfortunately for the possibilities of establishing a nuclear test ban, the political situation in late 1959 left much to be desired. Although on December 14, 1959, the Soviet Union apparently agreed to accept the Western position on a number of items such as staffing and the budget, these concessions were actually offered only as part of a package proposal unacceptable to the West. For inherent in the voting procedures and composition of the control commission proposed by the Soviet Union was a Russian veto over operations of the proposed commission.

The Attempt to Implement the Geneva System:
Phase IV

In the spring of 1960 American scientists, by use of chemical explosives in Project Cowboy, proved by actual experiments that the decoupling theory—contrary to Russian objections—was not speculative but quite valid. Furthermore, the experiments disclosed a significant new facet of decoupling which meant further substantial deterioration of the capabilities of the Geneva System. It was discovered that sizeable decoupling factors could be obtained with holes far smaller than those initially postulated by Latter's theory. For example, if a violator only required a decoupling factor of 30 to evade the detection system, as would be the case for relatively small explosions, i.e., 20 kilotons or less, he could achieve this with a hole of only $\frac{1}{10}$ (Bethe's estimate) to $\frac{1}{30}$ (Latter's estimate) of the optimum size needed to provide a decoupling factor of 300. This was a very important discovery in terms of the practicability of decoupling; the effort required to construct cavities could be greatly reduced, especially for the decoupling of small explosions.[16]

In addition to the discovery of this phenomenon of partial decoupling, the practicality of decoupling was further strengthened in April 1960 by the realization that the technology already existed in both the Soviet Union and the United States to make cavities which would provide substantial decoupling factors. These cavities could be constructed through the use of solution mining techniques in salt domes, and there are over 200 such salt domes in the United States and at least 100 in the Soviet Union. There are already about 300 cavities of large proportions in the American domes which are used for storage of petroleum products and for brine production. The Joint Committee on Atomic Energy has concluded on the basis of expert testimony that "it is feasible to construct cavities in salt domes of the required depth, size, and shape to decouple nuclear explosions up to about 100 kilotons."[17]

In spite of these technical setbacks to the Geneva System, political developments throughout the spring of 1960 gave considerable cause for optimism. President Eisenhower remained firmly convinced of the wisdom of a nuclear test ban and sought to make one final attempt to reach agreement with the Russians prior to the termination of his term of office. Specifically, the Administration decided to reiterate, in a new guise, its suggestion of April 13, 1959, that nuclear tests be outlawed gradually as the capabilities of the control system improve.

This new American initiative was translated into policy on February 11, 1960, when the Western powers offered a

[16] *ibid.*, pp. 183–85. On the other hand, more recent evidence suggests that the decoupling factor in salt formations is only 120 rather than the 300 calculated by Latter; *The New York Times*, December 19, 1961, p. 5.
[17] U.S. Congress, Joint Committee on Atomic Energy, *Report on Technical Aspects of Detection and Inspection Controls of a Nuclear Weapons Test Ban*, 86th Congress, 2nd Session, May 1960, p. 27.

treaty to the Soviet Union based on the Geneva System of 180 stations. Whereas the April 13th proposal in 1959 would have banned only atmospheric and oceanic tests, this treaty proposed to outlaw all detectable tests underground, in the atmosphere, in the oceans, and in those regions of space where effective detection could be exercised.

With respect to underground tests, the February 11th proposal would establish a "threshold" at 4.75 seismic magnitude above which testing would be outlawed and below which testing would be permissible. The threshold of 4.75 seismic magnitude was chosen because it corresponded to the Western evaluation of the detection and identification capabilities of the Geneva System, i.e., 90 per cent of the seismic events with a magnitude of 4.75 and above would be identifiable by seismic means alone; the remaining 10 per cent of the seismic events would be identifiable only through on-site inspection.

As a 4.75 seismic magnitude is equivalent (according to Western scientists, at least) to approximately a 20 kiloton nuclear explosion fully coupled, then, under the treaty, the United States or any other nation could legally test any weapon underground up to 20 kilotons. If the nation should choose to decouple the explosion and thus decrease the magnitude of the seismic reading, it could legally test weapons of a few hundred kilotons.

The Western proposal also provided for a joint research program including the use of nuclear explosives to improve underground detection. Then, as detection improved through research, the ban on testing would be broadened to include tests of lower seismic readings until all underground testing would be banned. Also, under this proposal the Soviet Union would have to agree to permit adequate inspection of the suspicious events recorded by the control system. Attempting to avoid conflict over the criteria to determine the initiation of on-site inspection, the United States accepted the notion of a quota system of yearly inspections as first proposed by Prime Minister Harold Macmillan. In his presentation of the proposal to the Soviet Union at Geneva, United States Ambassador James Wadsworth suggested as an adequate number 20 inspections a year within the USSR.

The suggestion of a quota of 20 inspections per year can be viewed both as a concession to the Soviet reluctance to permit inspection of its territory and as consistent with the oft-stated American position that the number of on-site inspections should bear an "appropriate relationship to scientific facts and detection capabilities." Whereas at the Geneva Conference of Experts the West appeared to take the position that inspection meant that *all* suspicious events were to be inspectable, under this proposal *only 30 per cent* of the predicted seventy annual suspicious events in the USSR of 4.75 seismic magnitude and above, i.e., 20 events annually, would be inspectable. Such a number apparently was chosen because it was believed by the Administration to be sufficiently large *to deter* evasion by the Soviet Union yet sufficiently small to be politically acceptable to the USSR.[18]

The immediate Russian reaction to the Western proposal was that it was "unacceptable." The Russians argued that it was a meaningless subterfuge which would only sanction recommencement of testing by the United States. The Soviet Union would accept nothing less than a total ban on all nuclear weapons tests, and on February 16 it countered with a proposal of its own. In return for a *total* ban on all tests, the Soviet Union would agree to permit the West to inspect a quota of events; it also proposed more liberal criteria for initiation of inspection. The size of the quota, the Russians held, was a "political" question for negotiation and not, as the United States claimed, a "scientific" one. By "political" the Russians apparently meant, as later events were to show, that the number of inspections should be very small and

[18] According to one knowledgeable informant, at no time have Western representatives held that *all* suspicious events in the USSR would have to be eligible for inspection. For a discussion of this matter see U.S. Congress, Joint Committee on Atomic Energy, *Hearings on Technical Aspects* . . . . , *op.cit.,* p. 75ff.

merely "symbolic" of good intentions of the signatories to the treaty.

Nevertheless, the Western Powers found the Soviet reaction to the Western proposal encouraging. As it was offered just three days after the emergence of France as a nuclear power, it was believed to be a reflection of the growing Soviet concern over the spread of nuclear weapons. Thus it was hoped that this fear would break down the Russian reluctance to permit meaningful inspection of its soil. The hopes generated in the West by the Soviet counterproposal of February 16 were strengthened further when on March 19, 1960, the Soviet Union was reported to have accepted President Eisenhower's proposal of February 11th. In return, the Russians said that the West would have to agree to a five-year moratorium on *all* tests not specifically covered by the treaty.

In one particular the Russian response was especially encouraging. For the first time in almost two years of negotiations, the Soviet Union apparently accepted "in principle" the threshold concept, i.e., "the implied acknowledgment that there do exist substantial problems in the detection and identification of underground explosions."[19] Perhaps now, it was reasoned, the Russians would agree to improvement of the Geneva System on the basis of further research and to a reasonable number of on-site inspections.

In much of the West the public response to these proposals and counterproposals was one of profound encouragement. The propaganda impact of the Russian move was great. *The New York Times* reported that the Russians at long last had accepted the American position with only minor reservations.[20] The headline in the *Washington Post* read: "Reds Accept Ike's Test Plan Provided Small Blasts Cease." The sentiments of the pro-ban scientists were well stated by Hans Bethe in a statement dated March 23: "The latest Russian

---

[19] U.S. Congress, Joint Committee on Atomic Energy, *Hearings on Technical Aspects . . . . , op.cit.,* p. 75.
[20] *The New York Times,* March 20, 1960, p. 1.

proposal made . . . [March 19] at Geneva is a major step forward in the negotiation for cessation of nuclear weapons tests. Provided the details can be satisfactorily cleared up, it seems to me a suitable basis for a treaty.

"The Russian proposal accepts all the points of the latest American proposal for a limited test ban. It adds to the American proposal only one important point, namely, that there should be a moratorium on small underground tests while the scientists of the nuclear powers do research on better methods for the detection of underground explosions."[21]

A strong discordant note was sounded by members of the Joint Committee on Atomic Energy. Senator Clinton Anderson, Chairman of the Joint Committee on Atomic Energy, labelled the Soviet proposal as "phony." At a later date Senator Albert Gore brought to light the unspoken conviction of many members of this committee whose task it is to be "watchdog of the atom." Senator Gore pointed out that the Russians, in their counterproposal of March 19th, had failed to accept the three crucial components in the President's proposal of February 11th: (a) on-site inspection—the Russians did not accept the United States suggested quota of 20 on-site inspections; (b) composition and procedure of the control commission—the United States and Russian positions were directly opposed in this important area; (c) seismic research—the USSR did not agree to a program of seismic research using *nuclear* devices.[22]

President Eisenhower's advisors were divided in their reaction. For example, Chairman McCone of the AEC was opposed to a long unpoliced moratorium. On the other hand, other advisors such as the Secretary of State and the President's Special Assistant for Science and Technology were far more favorably disposed toward the Russian proposal. Of perhaps decisive importance in reaching a decision was the

[21] *Congressional Record,* March 28, 1960, p. 6216.
[22] U.S. Congress, Joint Committee on Atomic Energy, *Hearings on Technical Aspects* . . . . , *op.cit.,* pp. 169–70.

President's strong desire for a nuclear test ban; this was reinforced by the urgings of Prime Minister Macmillan who had flown to Washington to discuss the matter with him.

On March 29, 1960, the two Western leaders agreed to join the Soviet Union in a moratorium on all underground nuclear explosions provided the Russians would actually submit to a system of international inspection and agree to a research program and to other details such as the West had proposed on February 11th.[23] The President told his news conference of the thinking that underlay his decision: "Now, they have come a long way since they said, 'Now we are ready to establish these mutual systems,' and so the very fact that they have made this concession means that they want to negotiate further; no question in my mind."[24] It must be noted that in taking this action the President abandoned the second qualification of his August 22, 1958 policy statement on a nuclear test ban, namely that the United States would continue its suspension of testing only if an *effective* inspection system were in existence.

The prospects for agreement on the main outlines of a treaty at the Paris summit meeting to be held in mid-May 1960, began to brighten. All that remained before a treaty could be signed was to work out the details. Bethe, in his March 23rd statement, expressed the conviction that the clearing up of these details would soon lead to the achievement of a first step toward disarmament: "Obviously the U.S. will have to seek further clarification of the Russian proposal. In particular, we must be assured that the Russians will permit an adequate number of on-site inspections, and liberal criteria for initiating them. Furthermore, the duration of the moratorium will have to be negotiated. But if these and other details can be satisfactorily cleared up the Russian proposal seems to me to give a good basis for an agree-

---

[23] *The New York Times,* March 30, 1960, p. 1. The subject of space tests was apparently left undecided.
[24] *The New York Times,* March 31, 1960, p. 14.

ment of test cessation at rather small risk. The risks in the continued and intensified arms race are immeasurably greater."[25]

The hope which Bethe shared with many others that the Paris summit conference could pave the way for the resolution of the issues preventing East-West agreement on a nuclear test ban proved to be short-lived. Nevertheless, the continuation of test ban negotiations despite the failure at Paris was believed to augur well for success in the future.

The belief that the test ban negotiations might yet succeed was reinforced by Russian concessions following the collapse of the Paris Conference. On July 26, 1960, for example, the Soviet Union offered to allow international inspection teams to carry out three on-site inspections annually on its territory. Although the West regarded this number as "grossly inadequate," it was accepted as an indication of Soviet willingness at least to negotiate the matter after 15 months of evading the inspection issue.[26]

Following the Paris debacle, however, President Eisenhower saw little reason to share this hope for the eventual success of the Geneva negotiations. By mid-summer of 1960 he believed that the prospects for a nuclear test ban were dead and that the United States ought not to continue its moratorium on nuclear testing.[27] Nevertheless, although the Pres-

---

[25] *Congressional Record,* March 28, 1960, p. 6217. The Subcommittee on Disarmament of the Senate Committee on Foreign Relations, under the chairmanship of Senator Hubert Humphrey has released an excellent summary of the twenty-five unsettled issues as of October 1960. U.S. Senate Subcommittee on Disarmament of the Committee on Foreign Relations, *Conference on the Discontinuance of Nuclear Weapons Tests— Analysis of Progress and Positions of the Participating Parties, October 1958–August 1960,* 86th Congress, 2nd Session, October, 1960.

[26] *The New York Times,* July 27, 1960, p. 1.

[27] This discussion is based on a letter to *The New York Times* by Lewis Strauss dated June 19, 1960. It will be remembered that in agreeing to negotiate a nuclear test ban in 1958, the U.S. had imposed upon itself a one year moratorium. This was extended on August 26, 1959 until December 29, 1959. From January 1, 1960 to the recommencement of American testing in September 1961, the U.S. subjected itself to a day-by-day moratorium.

ident told political intimates of his changed position on the testing issue, he did not believe that it would be proper to reverse his policy. In other words, President Eisenhower believed that he should not make a decision so late in his term which would by necessity bind his successor.

The Attempt to Implement the Geneva System:
Phase V

President Eisenhower's successor, John F. Kennedy, entered office with the conviction that while victory or defeat for the United States in the East-West struggle depended upon the success of American policy in the non-aligned areas of the world, human survival itself depended upon an East-West agreement to end the nuclear arms race. Desiring both to improve the American image with the neutral nations and to end the nuclear arms race, President Kennedy believed one more attempt to achieve a nuclear test ban ought to be pursued.

In an attempt to increase the probability of success in this last try for agreement, President Kennedy convened a panel of scientists under James Fisk to evaluate the American position and to recommend modifications in the American position which might break the existing deadlock between East and West. Out of Fisk's panel and the work of the President's Disarmament Advisor, John McCloy, came a draft treaty on April 18th and there followed many subsequent concessions including a reduction (from twenty to twelve) in the number of inspections of Soviet territory.[28]

At the same time, however, that the Kennedy Administration was trying to reach a reasonable accommodation with the Soviet Union, the latter was increasing its political demands. Whereas the Soviet Union had appeared to relax its veto demands in the late spring of 1959, in March 1961 it returned to its position of 1958 that it should exercise a

[28] *The New York Times,* May 30, 1961, p. 4.

veto over all substantive decisions of the control commission. The only change was that whereas the control commission proposed by the Russians in 1958 would have contained a British, an American, and a Russian member, the latest Soviet proposal based on the *troika* principle was for a three man council representative of East, West, and the neutral bloc; each council member would have the right of veto.[29]

A second major development in Soviet disarmament policy which was soon to confound President Kennedy's hopes for a nuclear test ban was Russian adoption of a policy which the United States had followed prior to August 22, 1958, i.e., that a nuclear test ban could not be an "isolated step." Whereas in 1957 the Soviet Union had argued that a test ban would be a good thing in itself, it now argued that a test ban was meaningless without "universal and complete disarmament."

Then, on September 1, 1961, the Soviet Union recommenced its testing with a program which was to include at least fifty atmospheric detonations and one explosion equivalent to approximately fifty-seven million tons of TNT. The Soviet lead was followed by the United States on September 15th when it began a modest series of underground tests. Neither side, however, felt impelled to withdraw from the test ban negotiations which continued at least in name.

There can be little doubt that the abrupt Soviet recommencement of testing and especially the intensity of the testing program came as a startling surprise to the Kennedy Administration and its scientific advisors. The most vital question which had to be answered at this point was whether

---

[29] *The New York Times,* March 22, 1961, p. 1. The Kennedy Administration did agree to the previous Russian demand for East-West parity on the control commission. The Kennedy Administration also agreed to reduce the U.S. demand from 21 to 19 control posts on Russian soil. The Russians have yet to agree to the establishment of one control post on their territory.

Other minor concessions of the Kennedy Administration were made with respect to the length of the moratorium and the research program to accompany a test ban treaty.

the self-imposed moratorium on American testing had given the Soviet Union an opportunity to overcome the American lead in nuclear weapons which had been evident to the Bethe Panel in 1958.

To make this determination, President Kennedy appointed a panel of experts under the chairmanship of Hans Bethe to analyze the debris from the Soviet explosions and to evaluate the extent of Soviet progress. As a result of its analysis, the panel of experts concluded that although the United States retained an over-all lead in nuclear weaponry, the Soviet Union had made important cuts into the American lead and might have surpassed the United States in certain categories of weapons. The available evidence pointed to the following examples of Soviet progress: (1) Reduction of weight-to-yield ratio; (2) increase in absolute yield of warheads; (3) reduction in size of fission trigger; and (4) proof testing of designs under combat conditions.[30]

As a consequence of this technical progress, the Russians were not only able to develop a deliverable fifty—and possibly a one-hundred—megaton missile warhead, but they also were able to reduce the size of less powerful warheads and to progress toward an anti-missile missile. Of equal significance is the fact that they made progress toward a "clean" or neutron bomb; whereas past Soviet nuclear explosions were 50 per cent fission in nature, the Soviets were able to reduce this to 20 per cent with respect to the whole test series and to 3 to 5 per cent with respect to the 57 megaton explosion of October 30, 1961.[31]

Against this evidence that the American three year moratorium on all testing and then on atmospheric testing had

[30] *The New York Times,* December 8, 1961, p. 1.

[31] *The New York Times,* December 17, 1961, p. 10E, and the *Washington Post,* December 15, 1961, p. A6. As of 1958, the United States had reduced the fission trigger to 5 per cent of the total yield of the explosion. If a neutron or clean bomb means a large nuclear explosion detonated high above the earth such that the released neutrons and not the resulting blast strike the earth, then the USSR have taken a major step toward its achievement.

enabled the Russians to make substantial progress at a time when United States progress had been retarded, President Kennedy had to weigh his desire for an arms control agreement and his desire to maintain the good opinion of the neutralist bloc. With these ends in mind, he accepted, in a speech to the United Nations on September 25, 1961, the Soviet goal of "general and complete" disarmament. The President stipulated, however, that each step toward this ultimate goal had to be subject to verification through international inspection and that the first step was an inspected nuclear test ban.[32]

The Administration recognized the error of adopting a self-imposed moratorium of weapons testing when there was no way of knowing what the Russians were doing. Never again, officials determined, would the United States permit the Soviet Union to employ arms control negotiations as a tactic to gain surreptitiously a nuclear weapons lead over the United States. The problem of the moment was, however, what the United States ought to do given the Russian advances, public opinion against American testing, and the need for arms control.

Few of the President's advisors argued that the United States should not continue its testing underground. The advisors were divided, however, on whether or not the Russian threat to overtake the American nuclear lead necessitated the recommencement of atmospheric tests. To some advisors such as Hans Bethe it seemed that the United States could maintain its lead through underground testing. The military, on the other hand, argued that the military requirements of the United States such as improved missile warheads and a warhead for an anti-missile missile required atmospheric testing.

Throughout the winter of 1961–1962, the President remained reluctant to resume atmospheric testing. Against the military

[32] See the President's speech as reported in *The New York Times,* September 26, 1961, p. 1.

argument he balanced the dangers of atmospheric fallout, the reaction of world public opinion, and the resultant acceleration of the nuclear arms race. It was becoming abundantly clear, however, that little progress could be made through underground testing and that the United States could not afford to retard its nuclear weapons program at the same time that the Russians might be secretly preparing yet another new test series.

Confronted with the danger of new Soviet tests which might shift the balance of power and with the need of the United States to test new weapons designs, on March 2, 1962 President Kennedy delivered an ultimatum to the Soviet Union through an address to the American people. The United States, the President stated, would recommence atmospheric testing in late April unless the Soviet Union at the forthcoming Geneva Disarmament Conference (March 14, 1962) would accept international inspection of a nuclear test ban, including inspection of the preparations for testing. With regret the President had concluded that a nuclear test ban agreement with the Russians as a first step toward disarmament had failed.

Conclusion

Paradoxical as it may seem, both the pro-ban and the anti-ban scientists believe they have been vindicated by developments in the test ban negotiations from 1958–1962. The two groups of scientists have drawn widely varying conclusions on the meaning of the three year negotiations and of their apparent failure. The underlying non-technical assumptions of both groups of scientists have been reinforced and the cleavage within the scientific community over nuclear weapons policy continues.

The pro-ban scientists believe that the Soviet Union initially and even until early 1961 sincerely desired a nuclear test ban. While the Russian intransigence on inspection was admittedly a major obstacle, these scientists believe this could have been overcome if the United States had negotiated in greater earnest. This the Eisenhower Administration was never able to do, these scientists point out, because it was divided within itself on the advisability of a nuclear test ban. Therefore, in the view of these scientists, the tragedy lies in the fact that the Kennedy Administration, with its whole-hearted support for the test ban, came to power too late. By the time it assumed power and began negotiations, the Kremlin had changed its priorities and was no longer interested in a nuclear test ban.

Undoubtedy these scientists would agree with the assessment of the failure to achieve a nuclear test ban treaty made by David Inglis in July 1961: "Two years ago the Soviets may have wanted a test ban. But our failure to resolve the divided opinion in our Government made it impossible to explore this possibility with effective negotiation."[33] In the meantime, they argue, the pressure of Communist China on Russia to break off the negotiations, the increasing desire of the Russian military for further weapons tests, and the Communists' appreciation of the inroads upon the Iron Curtain which inspection would make necessary have all contributed to a reorientation of Soviet policy.

Despite their criticisms, these scientists believe that the three year negotiations and the accompanying moratorium on testing have been to the American advantage. In the first place, the Soviet breaking of the moratorium shifted the onus for the failure of the negotiations onto the Russians. In the second place, scientists such as Hans Bethe have noted that the Russian recommencement of testing with megaton weapons justifies the belief that the continuing moratorium had favored the military position of the United States. These scientists reason that as "underground testing requires the development of entirely new methods to assess the results, and that these methods would be time-consuming and costly,"[34] the Russians must have concluded they could not

[33] *The New York Times,* July 9, 1961, p. E6.
[34] *The New York Times,* June 25, 1961, p. E7.

cancel the American lead in 1958 of 153 tests to Russia's 55 known tests through clandestine testing. Furthermore, the fact that the Soviets tested megaton weapons confirmed for these scientists their belief that the Soviet Union has little interest in tactical nuclear weapons and that American concern over Russian cheating with small explosions had caused an unnecessary loss of an opportunity to prevent the Russians from testing large scale weapons through agreement on a test ban.

To the pro-ban scientists, the nuclear test ban negotiations, like the Baruch Plan negotiations, have been a worthwhile experiment to test their belief that disarmament and arms control are feasible. Even though the experiment appears to have failed, these men believe that much has been learned from the experience and that the United States has a responsibility to absorb and build upon the basis of this experience in its continued attempts to achieve disarmament.

Perhaps the foremost lesson of the test ban negotiations as seen by the pro-ban scientists is that the United States asked the Russians for "a lot of inspection . . . for so little disarmament."[35] Thus, in direct conflict with their original view that a test ban would be a feasible first step to disarmament precisely because it would be relatively easy to inspect, many now argue that a more ambitious approach might succeed where a less ambitious plan had failed. This thesis was well expressed by Jerome Wiesner, President Kennedy's Special Assistant for Science and Technology: "The limited proposals [such as a nuclear test ban] require either the Soviet Union and its allies to accept more inspection than they are prepared to have without extensive disarmament, or the West to accept arms limitations with what to it appears to be inadequate inspection."[36]

[35] Hans Bethe, as reported in the *New York Herald Tribune,* October 31, 1961, p. 8.
[36] Jerome Wiesner, "Comprehensive Arms-Limitation Systems" in Donald Brennan, ed., *Arms Control, Disarmament, and National Security,* George Braziller, 1961, p. 199.

The solution to the problem of arms control is therefore, at least in the opinion of Wiesner and some other finite containment scientists, "to design a comprehensive arms-control system which commences with low-risk measures that can be carried out in the atmosphere of suspicion and fear, but which clearly leads to the ultimate objectives . . . . If there exists an agreed upon long-term goal, a plan for reaching it by means of a sequence of arms-limitation measures and a timetable for doing so, there will be an enormous interest in the ultimate objective and individual steps will not have to be as finely balanced as if they were likely to persist for all time."[37] In other words, the interest of both East and West in the achievement of the total arms control system would be sufficient to encourage one side to risk a temporary insecurity and to discourage the other side from taking advantage of the other's act of good faith.

It is argued that if East and West could agree upon a plan for "universal and general disarmament" with inspection of each step such as was proposed by President Kennedy on September 25, 1961, their mutual interest in the plan's success would make it easier for the Russians in the case of a nuclear test ban "to accept more inspection than they are prepared to have without extensive disarmament." Similarly, just as the Russians must put good faith in the Americans with respect to a nuclear test ban and its inspection requirements, successive steps will necessitate that at some point the West place trust in Russian good intentions. These scientists reason that only in this way can this terrible impasse be overcome.

Thus the finite containment scientists appear to have returned to a theme first expressed in the 1945 Franck Report that "only lack of mutual trust, and not lack of desire for agreement, can stand in the path of an efficient agreement for the prevention of nuclear warfare. The achievement of such an agreement will thus essentially depend on the integrity

[37] *ibid.*, pp. 200–201.

of intentions and readiness to sacrifice the necessary fraction of one's own sovereignty, by all parties to the agreement."[38]

In decided contrast to the moral drawn by the pro-ban scientists, and perhaps by the Kennedy Administration also, the anti-ban scientists see the test ban negotiations as a major defeat for the United States. They consider that the negotiations reveal a tragic history of the United States having permitted itself to be maneuvered into an untenable political position and having allowed the USSR to cut the American lead in nuclear weapons development. According to this view, the Soviet Union was never sincerely interested in an inspected suspension of nuclear tests. The purpose of the USSR, they argue, was (1) to maneuver the United States into a position where it would not dare test new weapons designs; (2) to give the Soviet Union an opportunity, through secret underground testing and then overt atmospheric testing, to catch up with the United States in nuclear arms; and (3) to outlaw through the pressure of world public opinion the use of the tactical nuclear weapons upon which these scientists believe the West must base its defense.

Furthermore, these scientists believe that the suspension of American testing after 1958 and the assumed Russian cheating which culminated in their open resumption of tests "very likely [gave] the Russians . . . a decisive advantage" over the Western powers and thus brought nuclear war that much closer.[39] These scientists fear that Russia may have nearly perfected an anti-missile missile which would give it an advantage in total war and also a "neutron" bomb which would enable them to win any limited wars. In the view of these scientists, the Russian fifty-seven megaton bomb cannot be discounted simply as a political maneuver; instead it represents progress whereby the Russians may have "tried

[38] See Chapter Two.
[39] Edward Teller, as reported in the *New York Herald Tribune*, October 13, 1961, p. 8.

out new principles developed in previous, undetected experiments."[40]

For these scientists, the lesson to be learned from this experience is not the need for a new approach to nuclear disarmament and arms control but the necessity of accelerated improvement in America's defense posture including an extensive shelter program and new nuclear weapons designs. They believe that any further weakening of America's defenses, especially with respect to nuclear weapons, will lead to war. Thus these scientists also return to a theme first expressed in 1945; this theme appeared in the report of the scientific advisory panel to the Interim Committee—that in the absence of an open world and world government, the threat to use nuclear weapons and not the elimination of these weapons is the key to lasting peace.

[40] *ibid.*

CHAPTER NINE

THE INTRA-SCIENTIFIC CONFLICT OVER A NUCLEAR TEST BAN: THE PROBLEM OF CONFLICTING EXPERTISE

THUS far this study has demonstrated (1) that the intra-scientific dispute over nuclear weapons policy is due in large part to the scientist's sense of social responsibility, (2) that in matters of high public policy political and technical factors are often too intertwined to be isolated from one another in the advice of scientists, and (3) that therefore the non-technical assumptions of scientists legitimately and frequently become part and parcel of the scientist's advice. As a consequence of these facts the problem of conflicting expertise has arisen. Even though scientists agree on the technical facts, their interpretations of the significance of the facts for public policy can be at great variance. This situation is well illustrated by the following juxtaposition of conflicting advice from two equally eminent scientists, Hans Bethe and Edward Teller, on the *technical feasibility* of a control system to monitor a nuclear test ban:

> HANS BETHE: "I believe, therefore, that it is *technically feasible* to devise a system of detection stations and inspections which give reasonable assurance against clandestine testing, with the possible exception of very small, decoupled tests."[1]

> EDWARD TELLER: "This is the impasse at which we find ourselves today. We can say simply, surely, and clearly that if we agree on test cessation today, we have no way of knowing whether the Russians are testing or not. *There are no technical methods* to police a test ban."[2]

[1] Hans Bethe, "The Case for Ending Nuclear Tests," *The Atlantic Monthly*, Vol. 206, No. 2, August 1960, p. 48. (Italics mine.)
[2] Edward Teller, "The Issue of Peace," Unpublished paper, p. 6. (Italics mine.)

Each scientist has interpreted the scientific facts according to his own political predispositions. To Bethe, a strong advocate of nuclear disarmament and a disbeliever in the strategy of limited nuclear war, the facts mean that the United States could have "reasonable assurance" that Russia is not cheating. To Teller, a strong advocate of a strategy of limited nuclear war and an opponent of nuclear disarmament, the same facts indicate that there are no "technical methods to police a test ban."

It is impossible to speak of either scientist as being correct because, although they speak as if the issue were solely a technical one, the questions dividing them actually are non-technical in nature. What constitutes "reasonable assurance" for Bethe does not do so for Teller. The significance of small, decoupled tests for Teller is not the same as it is for Bethe. Teller's judgment of the capacity of the "technical methods" to *deter* Russian cheating contrasts with Bethe's. Such differences in opinion cannot be settled by the methods of science. They rest on assumptions which are outside the scientific realm such as those concerning the military value of tactical nuclear weapons, the technological intentions of the Soviet Union, and the political desirability of disarmament.

This situation illlustrated by Teller's and Bethe's conflicting advice reveals how significant it is in the determination of public policy whether political leadership listens to the "scientific" advice of a Teller or of a Bethe. How perplexing, though, it must be for the politician or the administrator to decide between the advice of two such competent and convincing scientists. Yet choose between them he must, and the choices that have been made have had a profound and perhaps not always a beneficial consequence for American nuclear policy. For this reason, unless American political leadership learns to appreciate the nature of the problem of conflicting expertise, this problem will become an increasingly serious hazard to the formulation of an effective policy toward nuclear

weapons and toward other products of the scientific revolution as well.

The Nature of the Problem of Conflicting Expertise

That eminently capable natural scientists trained in a rigorous method of analysis can disagree so strongly with one another over a long period of time on the technical aspects of fallout or the technical feasibility of a nuclear test ban is quite amazing to most observers; since most people do not comprehend how such differences can honestly occur among scientists, the cause of the disagreement is often believed to be the dishonest behavior of one side or the other. It is frequently reasoned that scientists, if they are intellectually honest, should be able to agree on the precise issue dividing them, to define their terms, and thereby to reach agreement on the scientific facts as well as on the scientific unknowns.

Actually, the scientists, even in the midst of their most violent controversies, really have been able to attain a high degree of agreement on the scientific facts. Nevertheless, even though the intra-scientific debate over a subject such as the technical feasibility of a nuclear test ban leads to broad areas of consensus, the debate over the central issue continues without final resolution. The reasons for this dynamic quality of the debate must be understood if one is to appreciate the problem of conflicting expertise.

The dynamic or irresolvable nature of the intra-scientific debate over the technical feasibility of a nuclear test ban and other allegedly technical issues is due to certain elements which are found in the arguments of each side. This does not mean that these elements are deliberately contrived; if this were the case the professional conscience of the scientists would soon overcome what could be legitimately regarded as a case of intellectual dishonesty. Instead—and this accounts for the tenacity with which a scientist defends his position—the elements are implicit in the arguments of the scientist and the scientist himself is unaware of the manner in which they influence his judgment and his argument.

The first element which gives an irresolvable or *ad infinitum* quality to the intra-scientific conflict over the technical feasibility of a nuclear test ban is the conviction of each side that its position represents that of the "objective" scientist. Its scientific opponents, on the other hand, are regarded as having based their position on political grounds. As a consequence each side defines the issue in a manner which is consistent with its belief that it represents "objective" truth. Because neither side appreciates the fact that it has defined the issue in terms favorable to its own political position, the opposing positions cannot possibly define the issue in a way agreeable to one another.

Thus, for the proponents of a nuclear test ban, the debate seems to be between those scientists who like themselves believe that a test ban control system would have a good probability of catching, and consequently of deterring, a violator and those scientists who demand a control system with 100 per cent probability of catching a violator. As a 100 per cent certainty is scientifically impossible, the scientists favorable to a nuclear test ban tend to regard their opponents as deliberately distorting the scientific situation. They regard the demand of the anti-ban scientists for a foolproof system as an unfair and impossible technical requirement intended to destroy the confidence of the public in the efficacy of *any* inspection system.

The opponents of a test ban also define the issue of control in terms of their own political predispositions and thereby at times cast doubt on the integrity of the position of the pro-ban scientists. The debate, according to the anti-ban scientists, is between those scientists who like themselves believe that a test ban should have a good probability of catching a violator and those scientists who want a test ban based solely on trust in the Russians. Since the question of whether or not one can trust the Russians is a matter of political judgment, these scientists believe the proponents of a test ban are distorting their advice on technical feasibility in terms of their political judgment.

The second element of the debate which confounds any attempt to reach an overall consensus within the scientific community on the technical feasibility of a nuclear test ban is the impossibility of defining certain key terms such as "good probability." Both sides, unfortunately, use the term "good probability" as if it were clearly defined and were therefore a standard by which the adequacy of the control system could be measured. Yet, as has been indicated previously, each side actually defines the term in a way which supports its own political position. Thus whereas a 1-in-10,000 chance of a test being caught may seem to one observer to be a good probability, another person will not accept as a good probability anything with less than a 1-in-10 chance.

In actuality, one's view of how *good* the probability of catching a violator is and of how high it must be to deter violations of the control system is the composite of a number of judgments on certain qualitative matters: the probable efficacy of on-site inspection; the cleverness of the Russians and their determination to cheat; and the capabilities of a control system not yet in existence. For these reasons there is wide latitude for conflicting interpretations of the technical feasibility of the control system.

The problem of conflicting expertise is further complicated by the fact that the great number of unknowns, the unpredictable nature of scientific advance, and the latitude for qualitative judgments permit the continuous interjection into the debate of *ad hoc* hypotheses to reinforce a crumbling intellectual position. For example, when the anti-ban scientists introduced the decoupling theory and thus weakened the Geneva System, the pro-ban scientists countered with the argument that the theory was impractical even though valid theoretically and that non-technical means such as defectors and semi-technical means such as the possible observation of the large scale earth moving activities necessary to create a "big hole" would ensure Soviet compliance. A similar substitution and utilization of *ad hoc* hypotheses have been apparent as the anti-ban scientists have defended their basic position when it has been challenged by new findings of the pro-ban scientists.

As a consequence of these elements—opposed definitions of the issue at stake, differing definitions of qualitative terms, and the use of *ad hoc* hypotheses—the scientists are unable to devise a "crucial" experiment by which to resolve their differences. For this reason one cannot depend upon the methods of science to validate or invalidate the conflicting cases of the scientists; instead he must turn to an examination and evaluation of the non-technical assumptions which really divide the scientists. Fortunately, the debate over a nuclear test ban has brought these conflicting ideas to the surface. Following a discussion of the opposed technical positions of the scientists, therefore, this study will turn to a consideration of these underlying sources of disagreement.

The Conflict over the Technical Feasibility of a Nuclear Test Ban

THE POSITION OF THE PRO-BAN SCIENTISTS

Despite the extreme disappointment of the pro-ban scientists over the technical and political setbacks to a nuclear test ban, their basic attitude throughout the negotiations has been that, whatever the technical weaknesses in the Geneva System, they are minor when compared to the supreme political goals being sought. These scientists believe that establishment of control posts in the Soviet Union would be the first step toward disarmament, the opening of the Soviet Union, and the solution to the Nth country problem.

Until the breakdown in the test ban negotiations in September 1961 these scientists held that too much progress had been made toward the establishment of a nuclear test ban to permit the negotiations to fail. The concessions that they believed the Soviet Union had made indicated to them that the Russians were as interested in a test ban as was the United States. Therefore they felt that, until the United States had exhausted every possibility of overcoming the obstacles encountered since the Conference of Experts and all rays of hope had vanished, the test ban should be pursued with continued energy. As the Federation of American Scientists put it in July 1960: " . . . we do not know that our efforts have failed because of Soviet intransigence. Our efforts have not been complete enough for us to be confident that we would have found a basis for agreement if such existed. Our Government has not undertaken on a sufficient scale the hard work and intensive research that are necessary for an informal political judgment on specific arms control proposals."[3]

The pro-ban scientists have based their confidence that a nuclear test ban, whatever the apparent obstacles, would prove feasible upon the assumption that the Russians would not cheat because the incentive for the Soviets to cheat would be outweighed by their incentive to obey the ban. The interest of the Soviet Union in a successful nuclear test ban was believed to be far too great for the Russians to jeopardize the chance for the ban's success through nuclear testing in a covert fashion. Thus it was believed that even a remote possibility of discovery would deter the Soviet Union from any attempt to infringe on the test ban. As Russia was believed to be more interested in preventing the rise of new nuclear powers than it was in increasing its own armory of atomic weapons, these scientists argued that it would have a strong incentive not to wreck a nuclear test ban.[4] "I do not think," Hans Bethe has written, "the Russians intend to violate a treaty banning weapons tests; I do not think that the Russians could risk cheating, even if there is only a small likelihood of being detected. Even if we had no system of physical stations detecting

[3] Federation of American Scientists, Press Release, July 16, 1960.
[4] U.S. Congress, Joint Committee on Atomic Energy, *Hearings on Technical Aspects of Detection and Inspection Controls of a Nuclear Weapons Test Ban,* 86th Congress, 2nd Session, 1960, p. 184.

nuclear tests, the Russians would not risk having some defector tell us about a clandestine nuclear explosion. If there were such a defector telling us of a Russian violation, it would not be very difficult to find physical evidence of it. I believe that the Soviet Union, which is posing as a peaceloving nation, whether rightly or wrongly, simply cannot afford to be caught in a violation, and therefore I think that it will not try to cheat."[5]

On the basis of this conviction that the Soviet Union would not dare to attempt to cheat under a test ban, the pro-ban scientists made their requirements for a control system far less stringent than the capabilities of the system believed necessary by the anti-ban scientists. Consequently, unlike the latter group, they tended to discount the decoupling theory as providing a serious obstacle to the policing of the test ban. Bethe again summarized the pro-ban view of the decoupling theory with the observation that "we are all behaving like a bunch of lunatics to take any such thing as the big hole seriously . . . ."[6]

These scientists argued further that even if the Russians wanted to cheat and to carry out a new test program the "restrictions imposed by any detection system, even a very imperfect one, would greatly impede their progress and slow it down to a snail's pace."[7] They believed any covert testing program would be detected long before it had made any significant progress. There are, these scientists argued, a variety of methods by which covert testing would come to the attention of the control system other than through agreed technical methods, e.g., defectors, location of suspicious events in an aseismic region, evidence of decoupling cavities such as abnormal brine concentration in rivers, and repetitive suspicious seismic events in the same location.

[5] Hans Bethe, *op.cit.*, pp. 45–46.
[6] "The Test Ban and the Big Hole," *Scientific American*, Vol. 202, No. 6, June 1960, p. 81.
[7] U.S. Congress, Joint Committee on Atomic Energy, *Hearings on Technical Aspects . . . . ,* *op.cit.*, p. 184.

Among these informal policing methods the most potent, these scientists held, would be the fear of exposure through defectors. According to one pro-ban scientist, the Russians would be deterred from violation because cheating "places the whole Russian system at the mercy of any defector."[8] Another of these scientists believed that the "right of defection" ought to be incorporated into the treaty itself and thus "a citizen . . . would have a [legal] duty to report knowledge of secret testing to the International Control Commission."[9]

Furthermore, the pro-ban scientists argued that if the Soviet Union really desired to develop small yield tactical weapons they would not go to all the trouble of accepting the principle of control, of agreeing to a moratorium on small tests, and of pressing so hard for a nuclear test ban. As Hans Bethe put it: "If the Russians really want tactical nuclear weapons—that is, nuclear weapons of small yield—then the best thing for them to do would be to resume testing of such small weapons officially, exactly as was suggested in the original proposal by President Eisenhower on February 11. The fact that they asked instead for a moratorium on small tests indicates to me that they do not put much weight on development of these weapons."[10]

In the light of all these considerations the pro-ban scientists argued that the United States should take "positive" steps toward achievement of a nuclear test ban. They believed that American emphasis on obstacles to the test ban such as the decoupling theory was unhelpful and actually insulting to the Russians. Such an approach they maintained, was not creating the "mutual trust" necessary to build the peace.

[8] Quoted from testimony by Richard Roberts, *ibid.*, p. 217.
[9] Jay Orear, Letter to *The New York Times*, January 6, 1960. It is interesting to note that Edward Teller made the same suggestion in 1946 with respect to the Baruch Plan (see Chapter Three). Orear later expanded his suggestion; "A New Approach to Inspection," *Bull. Atom. Sci.*, Vol. 17, No. 3, March 1961, pp. 107–10.
[10] Hans Bethe, *op.cit.*, pp. 49–50.

Nevertheless, in order to decrease the unavoidable risk which would be inevitably inherent in a nuclear test ban, the pro-ban scientists have favored a vast research program in techniques of detection. They have argued, in the words of Hans Bethe, that "the next round [of scientific advance] ought to go to the detection rather than to the concealment."[11] In this connection the pro-ban scientists rightfully have pointed out that in the techniques of detecting and identifying underground nuclear explosions there were, and are, avenues of research as yet relatively unexplored that might reduce the possibility of evasion. Certainly the further study of the properties of the seismic waves generated by underground explosions and earthquakes could well discover positive criteria by which to distinguish underground nuclear explosions from earthquakes. Similarly, research on space testing might make it possible to detect nuclear explosions millions of miles distant from the earth.

Because their case for the technical feasibility of a system to police a nuclear test ban has been largely dependent upon the acquisition of new knowledge, these scientists have advocated an extensive program of seismic and other types of research. They have been resentful that efforts were made to discredit the Geneva System and that so little had been done to improve its capabilities. As a result of this dissatisfaction, much credit for the establishment of an extensive research program in detection methods (Project Vela) must be given to these pro-ban scientists. This is not to imply, however, that their scientist opponents have not also supported increased research on detection as well as on evasion techniques.

In addition to indicating possible future improvement of the Geneva System based on the acquisition of new knowledge, the pro-ban scientists have argued that a very substantial increase in its capabilities could be achieved within the bounds of available knowledge. Several proposals, all involving an

[11] U.S. Congress, Joint Committee on Atomic Energy, *Hearings on Technical Aspects* . . . . , *op.cit.*, p. 177.

increase in the number of control posts, have been put forward by these scientists. For example, Bethe proposed a network of about 200 robot seismograph stations to be placed in the seismic regions of the USSR. Such a network would be able to identify a high proportion of the earthquakes equivalent in yield to a 20 kiloton explosion even though it were fully decoupled. Bethe argued that the remaining unidentified events requiring on-site inspection would be of a manageable number.

Most scientists agree that Bethe's network would cancel the reduction which decoupling would cause in the capability of the Geneva System. They disagree, however, on the question of whether or not the network would reduce sufficiently the number of necessary on-site inspections. In fact, anti-ban scientists have argued to the contrary that the number of suspicious events requiring on-site inspections recorded by the Bethe-proposed network would be as high as 560 per year for seismic events which register two kilotons or more. Furthermore, the number of robot stations would have to be increased to 600 if the whole of Russia were to be covered, and the Russians have rather consistently rejected any such use of robot stations, as their use would undoubtedly open all of Russia to roving teams of foreign inspectors to insure that the robots were operating properly.

In summary, at least prior to the Soviet resumption of nuclear testing in September 1961, the pro-ban scientists believed not only that the Geneva System could be improved sufficiently to make it much more reliable but also that the sheer effort involved in carrying out a covert test program, the uncertainty of its results, and grave fear of "getting caught" would deter the Russians and any other nation from cheating. For the most part this conviction of the pro-ban scientists rested on their assessment of Russian motives and of the efficacy of the Geneva System. In addition, however, one senses in the case of many of these scientists a feeling that the internationalism of science itself would operate as a policing mechanism. Through personal contacts with Soviet scientists, professional gossip, and "defectors," it was believed that American scientists would be able to judge whether or not the Russians were cheating. In addition, Soviet scientists, acutely aware of the dangers of the nuclear arms race, would have presumably been able to resist the pressures of the Soviet military for continued testing; professional conscience and awareness of the dangers in the nuclear arms race would deter these Russian scientists from supporting illicit activities if they knew their American colleagues were equally restrained.

Indicative of this unspoken assumption of many pro-ban scientists that Soviet scientists could be depended upon either to prevent Russian cheating or to make such cheating known to their American colleagues was the following portion of an interview given by Harrison Brown of the California Institute of Technology:

Question: "Would there not be the risk that the Russians might cheat during [a moratorium on testing] . . . ?"
Answer: "They might, but I doubt that they would. First of all, I have talked with many top Russian scientists during the last 2 years and I believe that they seriously want an agreement. Further, should they cheat there would always be the possibility that we would find out. The disadvantages to them, should this happen, would probably outweigh the advantages to be gained from clandestine testing."
Question: "Is there any direct evidence that the Russians are testing now?"
Answer: "There is no evidence. Indeed, Russian scientists of my acquaintance are as incensed at American suggestions that they might be testing, as we were incensed by Communist charges during the Korean War that we were resorting to the use of bacteriological warfare."[12]

Although not all pro-ban scientists would go as far as Brown in his dependence upon the statements of Russian

[12] Interview of Harrison Brown by *Los Angeles Times,* December 25, 1960. See *Congressional Record,* January 30, 1961, p. A566.

scientists, the American scientists who favor a nuclear test ban tend to remain convinced that the international community of science is a force for peace. There is an implicit faith underlying many of their activities, such as the Pugwash Movement, that through appeals to Russian scientists the Soviet Union can be guided toward a more rational pattern of behavior. Many American scientists believe that a test ban would mean the most significant opening of Soviet society since the beginning of the Soviet thaw. In their view, it would give real substance to the belief that the international cooperation of scientists in both scientific and political areas constitutes a major factor working for peace.

THE POSITION OF THE ANTI-BAN SCIENTISTS

The anti-ban scientists have based their case against a nuclear test ban on the argument that the probability of detection of a test violation would be too small to overcome the powerful military incentive to advance nuclear technology through a covert test program. In the view of these scientists the technical situation was and continues to be such that the only guarantee against covert testing would be the good faith of the parties involved; and the past history of the relationship between East and West as well as the potentialities inherent in further weapons development have made "good faith" too weak a basis for a decision to enter into a test ban agreement. Furthermore, they have argued, the technical and political situation which has been unfavorable to the success of a test ban is likely to continue to exist for a long time to come. These scientists have believed that, unless the Soviet Union becomes an open society, programs of covert testing could progress almost as easily under a test ban as have the well publicized testing programs.

In contrast to Bethe's belief that a covert test program would have to proceed at a "snail's pace," the anti-ban scientists have made the case that truly effective test programs could be carried out. In fact, these scientists believe, such test programs could be carried out even under a surveillance system far more stringent than any proposed systems and even under Bethe's proposed radical revision of the Geneva System with its 200 robot stations.

To illustrate their argument, the anti-ban scientists have pointed out that "the basic principles of most new [nuclear weapons] designs could be developed at energy releases below a few tens of kilotons" and consequently could be tested underground with little fear of detection. "Outer space testing would permit the checking of the data at full scale." In summary, these scientists have argued that "if a violator can cheat with explosions up to 1 kiloton, he can do all the work necessary to the development of small tactical weapons. If he can cheat to 20 to 30 kilotons he can do a lot, but not all he would probably want to do in developing strategic weapons. If, in addition, he can cheat in outer space in the megaton range, the full range of energy releases are available to him to conduct a full-scale weapons development program."[13]

In both space and underground testing, according to these scientists, the probability of "getting caught" is far too low to be a sufficient deterrent to cheating. They have noted that nuclear weapons of extremely small yield could be decoupled so as to create very small seismic waves. Furthermore, even if these waves were detected and the event located, it would be extremely difficult, and more likely impossible, to discover and therefore identify an underground nuclear explosion by drilling for radioactive debris. Lastly, these scientists have pointed out that when one speaks of the *location* of a seismic event he means only the definition of a 100–400 square mile area within which the event occurred. The event cannot be pinpointed more accurately because the level of scientific knowledge about seismic waves' travel time is too scanty. Some of these scientists suggest that the

[13] U.S. Congress, Joint Committee on Atomic Energy, *Hearings on Technical Aspects* . . . . , *op.cit.*, p. 404.

earth's crust is so irregular that it will never be possible to locate events very accurately because of the variation in travel time of seismic waves with varying geographical locations.

The anti-ban scientists would undoubtedly agree with the assessment of the difficulty of identification through on-site inspection made by James Fisk at Technical Working Group II: Fisk told the delegates that a scientist would seek "an almost infinitesimal target, within an area of 200 square kilometers, for example. If it were ever the purpose of a violator to attempt to hide such an event the task of an on-site inspection group might be so great and so difficult that the probability of success would be close to zero."[14]

With respect to space testing, the anti-ban scientists have argued that whereas such testing is feasible as shown by Project Argus, a developed capability to detect space tests would not be possible for many years. In accordance with the conclusions of Technical Working Group I, even after the control system had been installed it would be extremely limited in capability. And even if suspicious events could be detected there would be no means by which they could be unquestionably identified as nuclear explosions or by which it could be proved which country had been the violator.

The anti-ban scientists grant that research in seismic and space detection methods would improve considerably the detection capabilities of the Geneva System. However, they have discounted the significance of such improvements for a number of reasons. Barring the discovery of a distinguishing and completely reliable characteristic for the seismic waves caused by nuclear explosions, the number of on-site inspections would have to be increased as improvement in detection capabilities increased, and then the strong opposition of the

---

[14] Conference on the Discontinuance of Nuclear Weapons Tests, Technical Working Group II, *Verbatim Transcript of Thirteenth Meeting,* December 10, 1959, p. 66.

Soviet Union to on-site inspection would be encountered. Also, these scientists maintain, it is rather meaningless to speak as Bethe does of 200 or 600 seismic stations in the USSR when that nation has not yet agreed, even in principle, to having 21 stations on its soil.

These scientists have believed that the possibilities for evasion due to the imbalance between the art of detection and the art of evasion will be increased by future research. They have argued that the knowledge of evasion techniques is at a very primitive level, and that research has been seriously pursued in only one area, namely that of decoupling through the use of large chemical explosions in earth cavities.

The anti-ban scientists have pointed out that a number of other possible evasion techniques exist which, like decoupling, will prove feasible. Three of the possibilities which have been outlined by these scientists illustrate their argument.

The first suggestion is to test underneath the floor of a large body of water, such as an ocean, sea, or large lake. This procedure would have many advantages for a nation desiring to evade a nuclear test ban. As shore-line areas tend to be highly seismic, the recording of a seismic event might not in itself cause suspicions to arise. Also, even if detected and suspected, the location of the event would be very imprecise and its identification extremely difficult because the area probably would not be surrounded by seismic control stations. Of decisive importance, the most crucial techniques of on-site inspection would not be available; the investigating team would be unable to depend upon aerial surveillance to reduce the suspicious area to manageable proportions for drilling purposes. As a result the investigating team would have few clues to guide it in selecting a spot to drill in an area hundreds of miles square. If an underwater shot were also decoupled, the difficulties of discovering the violation might be even more considerable. Ominously, they have pointed out, the most propitious place to carry out such a procedure would be the narrow Kamchatka Peninsula of Siberia and near Karafuto Island which have 60 per cent of Russia's earthquakes.[15]

A second possibility for an evasion technique which has been suggested by these scientists would be to mask the first motion of a nuclear explosion through the use of multiple explosions. As Edward Teller explained to the Joint Committee on Atomic Energy, it was conceivable, although admittedly difficult, that a series of nuclear explosions could be detonated in such a way that the explosions " . . . can wipe out the characteristic signature of a nuclear explosion."[16] As a result of such a procedure, Teller argued before the committee, the seismic events would pass unrecognized by the control posts.

These scientists also have discussed the possibility of another evasion technique, i.e., the employment in the decoupling cavity of certain materials that would absorb the energy from the explosion and consequently reduce the resulting seismic waves. Scientists in the anti-ban school believe that through the use of this technique the decoupling factor could be increased ten-fold; i.e., a 3000 kiloton explosion could be made to appear as a one kiloton explosion (300 present decoupling factor times ten).

Essentially, then, the position of the anti-ban scientists has been that evasion techniques could increase much faster than could methods for detecting nuclear explosions. Not only do they believe that detection through technical means would be infeasible but these men have felt that reliance upon non-technical means, such as defectors and espionage, could not remedy the situation. There have been, these scientists have pointed out, no Soviet defectors as yet who have told the West of the progress of Soviet nuclear research, and the West has been surprised by the first Soviet atomic bomb,

[15] For discussion on this evasion method see U.S. Congress, Joint Committee on Atomic Energy, *Hearings on Technical Aspects* . . . . , *op.cit.*, pp. 232–38.
[16] *ibid.*, p. 160.

by its first hydrogen bomb, by its first earth satellite and, they might add, by its resumption of testing in September 1961.

In response to the argument that one should not expect perfection in a system to police a test ban, these scientists have replied that there should at least be *some* chance of detecting a violation. They have argued that without an open world the probability of catching a violator would be negligible while the consequences of successful covert testing could be decisive for the outcome of the struggle between East and West.

The Underlying Sources of Intra-scientific Conflict

The task of elucidating the non-technical assumptions underlying the conflicting interpretations of the scientists concerning the technical feasibility of a nuclear test ban has actually been carried out largely by the scientists themselves. The confrontation of opposed scientists before Congressional committees and in print over the past few years has made quite obvious the non-technical assumptions which influence their presentation of the technical facts. While it often remains difficult to know where scientific agreement ends and political polemics begin, the discerning non-scientists can find exposed the non-technical views upon which the scientist bases his case.

Essentially the conflict among scientists with respect to a nuclear test ban has been due to differing views over the questions of Western military strategy, the motivation of the Soviet Union, and the political desirability of a nuclear test ban. While this distinction between military and political views is necessary for analytical purposes, those views are intricately joined in the thought of each set of protagonists.

CONFLICT OVER WESTERN STRATEGY

The strategic issue at stake in the struggle over a nuclear test ban has been whether or not limited nuclear warfare constitutes a feasible alternative to massive retaliation. Universally rejecting the doctrine of massive retaliation, most scientists have realized a need to substitute for it some form of military power as long as political tension and conflict characterize the world. Here, however, agreement ends and for a variety of reasons each side of the nuclear test ban issue has favored a different Western strategy. The opponents of the test ban have advocated a strategy based primarily upon the employment of tactical nuclear weapons; the proponents generally have argued for a strategy based solely upon conventional weapons.

The strategic view of the anti-ban scientists. The infinite containment or anti-ban scientists have accepted as basic the notion that "the idea of massive retaliation is impractical and immoral."[17] It is immoral because it involves the slaughter of innocent civilians; it is impractical in a world where the Soviet Union also possesses a long-range nuclear capability. The only reason for possession of megaton weapons is to deter their use by the Soviet Union. "We keep them," Teller and Latter write, "as a counterthreat against the danger that we . . . should be subjected to a devastating attack. . . . We believe that the role of nuclear weapons in a future war is by no means the killing of millions of civilians."[18]

In the view of these scientists, if the doctrine of massive retaliation is to be discarded, a new Western strategy must be developed which counters the apparent Communist military advantages of "central location, superiority in massive conventional weapons and in manpower, and finally, a political orientation which permits them to assume the initiative without any moral scruples."[19] Such an alternative strategy, these men have argued, can only be created through the development of a capability to wage limited nuclear war.

[17] Edward Teller, "On the Feasibility of Arms Control," Unpublished paper, May 6, 1960, p. 13.

[18] Edward Teller and Albert Latter, *Our Nuclear Future,* Criterion Books, 1958, pp. 141–42.

[19] Edward Teller, "On the Feasibility of Arms Control," *op.cit.,* p. 14.

"Tactical nuclear weapons could enable us to build up a counter force which would neutralize . . . Soviet advantages."[20] Even though the Russians might have these weapons as well, they would still lose their natural advantages. Indeed " . . . the main role of nuclear weapons might well be to disperse any striking force so that the resistance of people defending their homes can become decisive. Nuclear weapons may well become the answer to massed armies . . . ."[21] and thus serve to eliminate from political life the scourge of major war.

The development of tactical nuclear weapons, in the opinion of the infinite containment scientists, is bringing about a military revolution equal to the change wrought by the original atomic bomb. It will enable small mobile forces to be in possession of very great firepower and thus to accomplish the same ends which have required the employment of mass armies in earlier wars.[22] In addressing the Association of the United States Army on the need for a nuclearized army, Teller has argued: "I believe that this warfare of extreme mobility and dispersion will turn out to be the key to real military power."[23]

In the view of these anti-ban scientists, the argument that any use of nuclear weapons in limited war would lead to total nuclear war is a specious one. Wars, they have argued, are limited by their objectives and geography and not by their weaponry. For this reason they believe the use of nuclear weapons in limited war would not pose a danger of world wide nuclear conflagration. Indeed, according to these scientists " . . . the distinction between a nuclear weapon and a conventional weapon is the distinction between an effective weapon and an outmoded weapon."[24] These scientists believe

[20] *ibid.*
[21] Edward Teller and Albert Latter, *op.cit.,* p. 171.
[22] Edward Teller, "On the Feasibility of Arms Control," *op.cit.,* p. 14.
[23] Edward Teller, "The Impact of Nuclear Weapons on Future Organization," *Army Information Digest,* Vol. 12, January 1957, p. 18.
[24] Edward Teller, "The Nature of Nuclear Warfare," *Bull. Atom. Sci.,* Vol. 13, No. 5, May 1957, p. 162.

that the Russian argument that limited nuclear war will lead to total nuclear war is an attempt to forestall the development of a limited nuclear war capability by the West; the Russians, through a subtle threat of nuclear blackmail, seek to prevent the development of this strategy which is disadvantageous to the achievement of Russia's goals.

If the Western allies are to develop a strategy based on a capability for limited nuclear war, it requires "very special kinds of nuclear weapons which are hard to develop and harder to perfect."[25] The development of these weapons is at a very primitive level. As the anti-ban scientists believe that these weapons could greatly contribute to the stabilization of the world by giving the West an alternative to massive retaliation, they believe there is an urgent need to continue the development of tactical nuclear arms.

The infinite containment school has argued further that continued testing will have three major effects on nuclear weapons technology. Firstly, testing will improve the economy and the performance of present weapons; deliverability could be improved; and yield could be increased. Secondly, continued testing will enable the West to develop an arsenal of the "clean," small yield nuclear weapons it needs to counterbalance Soviet conventional strengths. Principal among these military requirements are radical new designs for battlefield weapons; other nuclear weapons which these scientists include in any list of possible advances through continued research are an anti-missile missile, an anti-submarine depth charge, and improved missile warheads.

The third and most important argument for continued research, according to the infinite containment scientists, lies in the future. These scientists believe that nuclear weapons research stands at the threshold of a major breakthrough which will greatly affect all warfare. For example, Freeman Dyson, who is not himself a member of the infinite containment school, wrote in April 1960: "I believe that radically

[25] Edward Teller and Albert Latter, *op.cit.*, p. 142.

new kinds of nuclear weapons are technically possible, that the military and political effects of such weapons would be important, and that the development of such weapons can hardly be arrested by any means less drastic than international control of all nuclear operations."[26]

In particular Dyson had in mind what has come to be called the "neutron" bomb. It was the possibility of such a development which Teller had in mind in 1957 when he advocated continued testing to perfect a "clean" bomb. Apparently agreeing with Teller, Dyson has written: "A fission-free bomb, containing a small quantity of heavy hydrogen and no fissionable metal, is logically the third major step in weapon development after the existing fission and hydrogen bombs."[27] Such a neutron bomb would release a lethal ray of neutrons in all directions. There would, however, be little accompanying destruction caused by blast effect; furthermore, although this fissionless warhead would not be completely clean with respect to radioactive fallout, it would be "enormously less" dirty than existing nuclear weapons. Such a weapon, these scientists believe, would be an ideal anti-personnel weapon for limited war. In addition, some reason, its neutron ray might make it the ideal warhead for an anti-missile missile.

These scientists have argued that the exploitation of these

[26] Freeman Dyson, "The Future Development of Nuclear Weapons," *Foreign Affairs*, Vol. 38, No. 3, April 1960, p. 457. For Dyson's latest views on the neutron bomb see *Bull. Atom. Sci.*, Vol. 17, No. 7, September 1961, pp. 271–72.

[27] Freeman Dyson, "The Future Development of Nuclear Weapons," *op. cit.*, p. 458. His statement is the most authoritative and, in fact, the only one on the possibility of a "neutron" bomb by a scientist acquainted with the American nuclear weapons program. Scientists have been exceedingly uncommunicative on this subject. The only other statements of substance have been made by Senator Thomas Dodd. As a consequence the exact nature of the weapon at issue when someone refers to "neutron bomb" can mean many things in addition to a fission-free bomb. The only available technical discussion of these various possibilities known to the writer is the following by a British scientist: Nigel Calder, "Notes on the 'Neutron Bomb,'" *New Scientist*, Vol. 11, No. 244, July 20, 1961, pp. 145–47.

potential breakthroughs would revolutionize ground warfare. Nuclear weapons with a yield from one ton to one kiloton TNT equivalent would give small tactical units immense firepower; furthermore, this firepower could be employed without fear of indiscriminate destruction in densely populated areas like Western Europe. The infinite containment scientists believe that through the exploitation of these possibilities for the development of tactical nuclear weapons the outbreak of limited war could be deterred in the same fashion that strategic strength deters total nuclear war.

The military significance of these potential nuclear developments is so important, these scientists have argued, that the Soviet Union would have great incentive to continue work in the area, regardless of any test ban; and the West cannot risk unilateral Soviet development of such new weapons. If the Soviet Union should be the first to develop fission-free or "clean" nuclear weapons, it would secure a military advantage which would enable it to dominate future limited conflicts with the West. In this connection Dyson has noted that Russian physics journals indicate that their physicists have been working in this area since at least 1952.[28]

The theoretical possibility of a "clean" nuclear warhead well illustrates the primary argument of the infinite containment school for continued research, development, and testing of nuclear weapons. This argument is that the science and technology of atomic energy are very young. The attempt to arrest them in the immediate future would be comparable to the prohibition of the further development of chemical energy after the introduction of gunpowder in Europe. The possible implications of atomic energy for future warfare are no less revolutionary than were those of gunpowder when it was first introduced.

The strategic view of the pro-ban scientists. The pro-ban scientists, on the other hand, reject as infeasible the basing

---

[28] Freeman Dyson, "The Future Development of Nuclear Weapons," *op.cit.,* p. 459.

of Western strategy on the use of atomic weapons in limited war. Atomic weapons, according to their view, are by their very nature weapons of total war. Whereas in the years following the outbreak of the Korean War many of these same scientists advocated that the West develop tactical nuclear weapons, they came to the conclusion after the arrival of nuclear parity that tactical nuclear weapons were not to the advantage of the West; furthermore, their use might trigger the widening of any limited war into total nuclear war. David Inglis, a spokesman for this group, has written in opposition to the sharing of atomic weapons information with America's European allies that "the policy . . . is based on an unproved and essentially unprovable proposition (probably a myth) that a 'small' nuclear war could be fought in Europe without very great danger of growing into a 'big' nuclear war that must be avoided."[29]

The only reason for maintaining *any type* of nuclear weapon in one's arsenal is, in the view of the pro-ban scientists, to deter their use by one's opponents. The present Western capabilities in nuclear technology, are, these scientists believe, more than sufficient to achieve this goal. Accordingly, there is little need to continue the development of weapons that the West dare not use except as a last resort. Also, these scientists have argued, the United States could continue laboratory research on nuclear weapons and be prepared if the Soviet Union should unilaterally resume testing.

The pro-ban school believes that the West must maintain sufficient conventional military power to deter Communist limited aggression, especially in Europe. It fears the possibility that the West might become so dependent upon nuclear weapons that it could not defend itself against limited advances by Communist satellite nations except by means which might result in a nuclear holocaust. Inglis, in the letter referred to above, stated the concern of these scientists that

[29] Letter to the editor, *The New York Times,* July 22, 1959, p. 26.

an emphasis on nuclear weapons among America's European allies as desired by the infinite containment scientists would make likely " . . . the degeneration of NATO capabilities to wage a non-nuclear war against the Eastern satellites."[30] If such an eventuality were to occur, the West could not meet limited encroachment by Communist conventional forces without a major nuclear war. In time this could lead to the piecemeal capitulation of Europe and the rest of the Free World.

These scientists deny that limited nuclear warfare would be to the advantage of the West. They argue that in limited nuclear war the nation with the greatest reserve forces would be the winner, as this type of war would consume far more troops than conventional war. In addition, tactical atomic weapons could be used by the Communists to destroy port facilities, to interdict supply lines, and to bomb American overseas bases. The high attrition rate, the vulnerability of Western supply lines, and the political costs of this type of war are viewed by these scientists as but three cogent reasons against dependence upon tactical nuclear weapons. However, even if feasible, these scientists point out that one result of a limited nuclear war might be the obliteration of the country the West is seeking to defend and that such a consequence might well make other Communist threatened nations more susceptible to nuclear blackmail.

These scientists base their case against the feasibility of a limited nuclear war on the fact that whereas there is a clear qualitative distinction between nuclear and conventional weapons—such as the presence or absence of radioactivity—which provides an upper limit on the war which both sides can observe and know that the other is also observing, this would not be true in the case of "limited" nuclear war. In the latter type of war, under the pressures and excitement of battle, there could be a gradual acceleration in the size of the weapons employed. As a mutually understood upper limit would be lacking, a nation might be

[30] *ibid.*

forced against its will to increase the magnitude of nuclear weapons used until the limited war actually became total.

In addition, these scientists argue, the proponents of limited nuclear war fail to recognize the possible political advantages which would accrue to the Communists if the primary Western response to Communist conventional aggression were with tactical nuclear weapons. While technically the only difference between nuclear and conventional weapons is one of their relative efficiencies, in political terms there is a considerable difference, at least in the eyes of the great majority of the world's peoples. The use of nuclear weapons to counter a limited Communist aggression undertaken with conventional weapons would carry a high political cost for the West. In such a case there is little doubt that the Communists would find many ways to utilize to their advantage the universal fear of all nuclear weapons.

For all these reasons, then, a major strategic purpose of a test ban, in the view of at least some pro-ban scientists, would be to emphasize the fact that nuclear weapons *are* different. A test ban would serve notice to all that these weapons are too dangerous to be employed. In effect, they would become outlawed by each nation for reasons of self-interest. No nation, this argument goes, would seek to gain advantage in limited war by the employment of nuclear weapons at the risk of plunging the world into nuclear conflagration.[31]

The proponents of a test ban have argued further that its cost in potential military capabilities would be negligible. Nuclear technology, they believe, has reached a point of relative stagnation. Any gain such as the proposed neutron

---

[31] For an excellent presentation of this argument see Donald G. Brennan and Morton Halperin, "Policy Considerations of a Nuclear Test Ban," in Donald G. Brennan, ed., *Arms Control, Disarmament and National Security*, George Braziller, 1961, pp. 234–66. Brennan is a physicist; Halperin is a political scientist. Whether or not one agrees with the conclusions of their analysis, it does indicate the benefits to be derived from this type of cooperation between natural and social scientists.

bomb which might be made through continued testing would be minor compared to the political-strategic advantages of a test ban. The beneficial results which a test ban would bring make it a course of action far preferable to opposition to a test ban based on the speculation that nuclear science is on the verge of a major breakthrough.

In the view of these scientists, the so-called "clean" or neutron bomb is technically and militarily questionable. The high temperatures required to bring about the fusion of hydrogen nuclei, these scientists argue, are unlikely to be created by conventional means such as high explosives. Even though theoretically possible, such a development is at least many years in the future. And, even if it could be perfected, they question the possible military significance of the weapon.

Furthermore these scientists are convinced that further progress in nuclear technology would not be to the advantage of the United States—nor to the Soviet Union for that matter. Whereas the infinite containment scientists look with favor upon the possibility of a research breakthrough which would, in their estimation, give a military advantage to the United States, the pro-ban scientists disparage the effects of new advances in nuclear weaponry. The advance of nuclear weapons technology by further testing would only simplify the state of atomic art and thus make it easier for smaller nations to become nuclear powers. Such an unwanted development would result if the present nuclear powers proved the feasibility of making a neutron bomb which does not require costly fissionable materials.

In fact the dubious nature of any advantage accruing to either the United States or the USSR through further testing has constituted the basis for the hope of these scientists that a test ban would be observed. Whereas the loss which would be suffered by a nation caught cheating on a ban would be known, the possible gain through covert testing would be unknown. It has been argued that, given these two alternatives, the Russian incentive to test new concepts and thus to infringe on the ban would be small.

CONFLICT OVER THE POLITICAL DESIRABILITY
OF A NUCLEAR TEST BAN

The underlying difference among the scientists on the political desirability of a nuclear test ban rests on contrasting assumptions concerning the nature of the Soviet Union, of the nuclear arms race, and of international politics. An understanding of these differing assumptions is therefore essential if the conflict among scientists over nuclear weapons is to be comprehended.

The view of the pro-ban scientists. These scientists believe that the Soviet Union is characterized by an internal conflict between an irrational political heritage and the modern rational forces represented by Soviet science. They believe the refusal of the USSR to see the inherent danger in the nuclear arms race and to seek meaningful measures to end it to be due to the continued supremacy of vestigial irrational forces such as communist ideology and the Russian military services. The pressing need, in the view of these scientists, is therefore to strengthen the forces within the Soviet society which are working toward a rational solution to the problem of atomic energy. A fundamental political purpose of a nuclear test ban would be to accomplish this goal. These scientists believe that unless this occurs there is no hope for the world, as the rapid advancement of scientific technology is incompatible with an irrationally ordered world.

The view of these scientists concerning the nature of current international politics reinforces their conviction of the nature of Soviet society. They believe that the destructiveness of modern weapons has ushered in a new era of history where the old rule of power politics that disarmament must follow political settlement no longer applies. Today, in the words of the Franck Report, as all nations "shudder at the possibility of a sudden disintegration" and have a mutual interest in self-preservation, "only lack of mutual trust . . . can stand in the path of an efficient agreement for the prevention of nuclear warfare."

A number of corollaries flow from this premise; whereas in the past disarmament was presumed to be dependent upon prior guarantees of security and the settlement of political disputes, now the overwhelming interest which all nations are believed to have in nuclear disarmament has reversed this rule. For this reason disarmament can presently be viewed primarily as an autonomous problem which can be solved wholly or at least in part by technical means. Thus the pro-ban scientists believe that the criterion of technical feasibility could become the *modus operandi* in at least one area of international affairs and, in this way mankind would, in time, rid itself of the instruments of war and thus begin to substitute a more reasonable method than war for the settlement of international disputes.

The scientists who favor a nuclear test ban have approached it as an experiment by which to test their theory that mutual distrust is the major barrier to general disarmament; if the test ban could be agreed to and succeed it would prove to the nations that the criterion of technical feasibility in international negotiations opens the way for the far-reaching disarmament agreements they all desire. As one test ban proponent, Bernard Feld, has put it, "although such measures [to police a test ban] can hardly be said to constitute serious disarmament,[32] the difficulties are in many respects characteristic of the types of technical problems which will be encountered in future arms-limitation negotiations."[33] Or, in the words of Hans Bethe, "the main importance of our negotiations on the test cessation agreement comes . . . not from this agreement itself, important as it is, but from further agreements which must follow."[34]

[32] Feld's footnote, "This fact may diminish their significance, but not their importance."

[33] Bernard Feld, "Inspection Techniques of Arms Control," *Daedalus* (*Arms Control* Issue), Vol. 89, No. 4, Fall 1960, p. 862.

[34] Hans Bethe, *op.cit.*, p. 51.

It has been believed that through their experience in making a test ban work the nations would overcome the fears and insecurities which supposedly prevent them from realizing their mutual interest in the international control of atomic energy. Then, through technical measures such as those detailed in Seymour Melman's *Inspection for Disarmament*,[35] the nuclear powers could proceed step by step to disarm themselves.

As they believe that irrational and "traditional" thinking still predominates in the governing councils of nations, these scientists have assumed a responsibility to hasten mankind's achievement of a social consciousness appropriate to the facts of the atomic age. Herein lies the meaning of the oft-repeated remark that the most important point in achieving a nuclear test ban is a political one, since a nuclear test ban would break the pattern of traditionalist thinking that prevails among men and would be the first step to a new "political atmosphere in which . . . the use of highly destructive weapons becomes increasingly an unthinkable, inhuman act."[36]

The progress which these scientists believed the Geneva negotiations had made toward the achievement of a test ban confirmed, for a time at least, their conviction that a ban represented an opportunity to bring about fundamental changes, particularly in the supposedly irrational behavior of the Soviet Union. For example, this appears to account for Bethe's reluctance to have the technical difficulties of the Geneva System stressed too much; he feared that emphasis on evasion and cheating might insult the Russians and cause them to regress to what some would call an irrational position. It was no doubt this concern which led Bethe to write in August 1960: "I had the doubtful honor of presenting the theory of the big hole [decoupling] to the

---

[35] Seymour Melman, ed., *Inspection for Disarmament,* Columbia University Press, 1958.
[36] Seymour Melman, "The Political Implications of Inspection for Disarmament," *Journal of International Affairs,* Vol. 13, No. 1, Winter 1959, p. 44.

Russians in Geneva in November, 1959. I felt deeply embarrassed in so doing, because it implied that we considered the Russians capable of cheating on a massive scale. I think that they would have been quite justified if they had considered this an insult and had walked out of the negotiations in disgust."[37]

Within the United States the test ban advocates believe that they have succeeded in accomplishing a major political breakthrough. For the first time since 1946 they feel that they have been victorious over the internal forces within the American government which have opposed any attempts to end the arms race. Although the way in which this was achieved was in part through Pauling's appeal to emotion and fear, they believe a new atmosphere has been created in which rational solutions can be found and will be met with favor.

In addition to being a first step toward disarmament and a moderating influence in the Cold War, the scientist-proponents of a nuclear test ban have tended to argue that it would be the solution to the Nth country problem. However, there are varying ideas of the way in which a test ban would arrest the spread of nuclear weapons to other nations.[38]

The most popular or prevalent view among these scientists—and within both the Eisenhower and Kennedy Administrations as well—of the way in which a nuclear test ban would limit the number of nuclear powers is that the force of world public opinion would act as a restraint. Thus, when asked by a United States Senator to explain how the ban would operate, John McCone, then chairman of the AEC, replied that "the thinking is . . . that a combination of logic and world opinion would develop adherence

[37] Hans Bethe, *op.cit.*, p. 46.
[38] For a sobering discussion of the Nth country problem see *1970 Without Arms Control: Implications of Modern Weapons Technology*, Report of the Special Project Committee on Security Through Arms Control, National Planning Association, Washington, 1958.

by other countries."[39] Essentially the same point was made to another Senator by John McCloy, who at the time was President Kennedy's disarmament advisor.[40]

A far more sophisticated view has been presented by a natural scientist working in conjunction with a political scientist.[41] In the view of Donald Brennan and Morton Halperin, a nuclear test ban would symbolize to the world the intention of the Great Powers not to use nuclear weapons in limited war. The ban, by serving to emphasize the danger of such weapons, would discourage the present have-not powers from investing their scarce resources in nuclear armament.

The third suggestion for the manner in which a nuclear test ban could solve the Nth country problem has been made by Herman Kahn, although he himself does not necessarily subscribe to it. This is the notion that a nuclear test ban could be enforced by the nuclear powers. In effect a ban would constitute a Pax Russo-Americana, and Kahn as well as his reader is aware of the improbability of such a political realignment of the nuclear powers.

In the eyes of the proponents of a nuclear test ban continued testing offers no solution to the problem of atomic weapons; a test ban at least holds a promise of reward in this area of threatening disaster. In the last analysis, these men would argue, one has to take a chance—an act of faith—which would enable mankind to get ahead of the nuclear arms race. In this light they have asserted that a test ban is a calculated risk which is well worth taking.

The view of the anti-ban scientists. In contrast to their scientific opponents, these scientists believe that a nuclear test ban would be neither a step toward disarmament, a

---

[39] U.S. Senate, Committee on Foreign Relations, *Hearings Before a Subcommittee on the Geneva Test Ban Negotiations,* 86th Congress, 1st Session, 1959, p. 16.

[40] Letter from John McCloy to Senator Stuart Symington. See *Congressional Record,* June 12, 1961, p. 9275.

[41] See Donald Brennan and Morton Halperin, *op. cit.,* pp. 234–66.

means to lessen the aggressive nature of Soviet policy, nor a solution to the Nth country problem. Indeed, they argue, rather than solving these problems a nuclear test ban would bring dire consequences.

The real danger, these scientists believe, is that the West will be lulled into accepting an agreement which incorporates a policing system inadequate to prevent covert weapons advancement. They believe that the opportunity and the incentive to test will always be so great that the arms race can only be arrested by a radical change in the degree of openness in the world. "Once we have accomplished that," Teller told a Congressional committee, "then the way toward disarmament and toward peace lies open before us. Without it, however, we are just inventing one method after another how to fool ourselves."[42] In turn, these scientists believe, an open world must await the settlement of the political issues dividing East and West and the "withering away" of closed totalitarian societies.

Specifically, these scientists expect that a test ban would create the illusion that all was well with the world and that an atmosphere of complacency would then envelop the West. If the West let down its guard, made far-reaching concessions to the Russians, and disarmed itself, the Communists would be emboldened to seek their objectives through military aggression. At the very least, in the opinion of these scientists, a test ban would deprive the West of the major strategic advantage it now has over the Soviet Union. Not only would a ban prevent the development of the additional new varieties of nuclear weapons needed by the West, but it would cast an aura of illegitimacy about all nuclear weapons that might make them too immoral or too dangerous to use as a counter to Soviet conventional military power. In fact, these scientists argue, the main purpose of the Soviet Union in the nuclear test negotiations actually has

[42] U.S. Congress, Joint Committee on Atomic Energy, *Hearings on Technical Aspects* . . . . , *op.cit.*, p. 167.

been to outlaw the nuclear weapons upon which the United States has and must base its defense.

These scientists maintain that the hope that a test ban would open Soviet society is entirely illusory. Even though one were to agree with the pro-ban scientists that the Russian stress on secrecy and fear of espionage is in part "irrational," these scientists would still argue that the Russians are quite thoroughly convinced that it is more in their interest to maintain territorial secrecy as a strategic asset than to permit the establishment of an inspection system which would decrease significantly the extent of this security curtain. For the Russians, secrecy has become a major element in the balance of power as it protects their missile sites from discovery and consequently protects them from destruction. In fact, they believe that secrecy may be of such significance as to give the Soviet Union the strategic edge over the United States. The United States is only deceiving itself, these scientists contend, in believing that the Soviet Union will be induced by the promise of a nuclear test ban to open its territory and expose the whereabouts of its missiles.

The impact of a prolonged nuclear test ban on the American nuclear weapons program would be disastrous, these scientists further contend. Contrary to the argument of the pro-ban scientists that research could continue, these scientists point out that the continuation of such research is dependent upon testing just as other types of scientific research are dependent upon laboratory experimentation. Furthermore, America's successful weapons teams would break up as scientists left Los Alamos and Livermore for more interesting and socially approved work. In contrast, the Soviet Union could force its scientists to continue research and to search for potential breakthroughs which would make covert testing or a repudiation of the test ban worthwhile.

As for the Nth country problem these scientists argue that the diffusion and utilization of knowledge is far too rapid to be controlled except by a world government. Regrettable as it may be, this fact is of the essence of the scientific revolution which is making scientific knowledge available to all nations; the United States through its own generosity in the Atoms for Peace Program and in technical assistance programs has actually contributed to this process.

As a matter of fact, these scientists believe, the only nations which might be deterred by a nuclear test ban from testing nuclear weapons would be the nations of the democratic West. And even the belief that these nations would be deterred can be questioned. The constant threats of Soviet nuclear blackmail and general world insecurity are powerful incentives for have-not nations to develop their own nuclear capabilities. In addition to these factors the desires for national grandeur and an equal voice in international councils offer further motivation to possess nuclear arms.

Fortunately, these scientists argue, the next nations to become nuclear powers will be the European allies of the United States, and this is necessary for the security of the West. Furthermore, the United States through cooperation with them can control the dissemination of nuclear weapons to its NATO allies in a manner which will lessen the anarchistic spread of nuclear capabilities. Perhaps, these scientists hopefully reason, this experience will teach the world how to control nuclear weapons.

With respect to the enforcement of a nuclear test ban, these scientists ask how public opinion could deter a nation from testing. Even if one grants that a nation would not dare defy world public opinion—an assumption made highly questionable by the Russian recommencement of testing in September 1961—there must be some way to assign responsibility for a test which might be detected. However, since a test program could be carried out at high altitudes, in space, and in the oceans without any evidence by which to determine responsibility if the test were detected, then the guilty party could easily accuse other nuclear powers of having been the violators of a ban.

The events of the twentieth century have caused the infinite containment scientists to be especially concerned with the question of whether or not democracy has the determination and skill to meet the challenge of despotism. Prior to both World Wars, these scientists point out, the democracies failed to build their strength but instead they listened to talk of disarmament and were lulled to sleep. There is a fear among these scientists that political leaders cannot negotiate sound arms control agreements and at the same time call forth from the Western peoples the effort required if the West is to survive and war is to be deterred.

Instead of seeking to eliminate nuclear weapons through disarmament, these scientists believe that man must solve the problem of war itself. Modern science, including atomic weapons, seems to these scientists to have become a major force for peace. In the first place, man has reached that point in his technological evolution where he can gain more through scientific cooperation than through international conflict; through such cooperation man can build a world community where war would be unthinkable. In the second place, science has provided the threat of nuclear destruction as a deterrent to war. Just as the threat of strategic bombing deters total war, tactical nuclear weapons can be made to deter limited war. Thus, echoing the advice of the science advisory panel to the Interim Committee in 1945, these scientists believe that the military exploitation of the atom holds greater promise for the prevention of war than the benefit to man of the possible elimination of nuclear weapons.

Conclusion

The problem of conflicting expertise, as exemplified by the debate over the technical feasibility of a nuclear test ban, is one which in a sense has always existed for the public and for the policy maker. However, the increase in the complexity of the issues facing society and in their importance has accentuated the problem. Of great significance in complicating this problem is the fact that scientists believe they have a social responsibility to assist society in solving the problems created by scientific advance. This new social commitment of scientists has made it increasingly difficult for the scientist and the non-scientist to separate the expert's advice concerning *what is* from that concerned with *what ought to be*.

In no area does the problem of conflicting expertise present a more grave challenge to policy-making than in the vital area of national policy toward nuclear weapons. Even more discouraging is the fact that the complexity of the problem has increased through the years. At the time of the Baruch Plan the political assumptions of the scientists had recently been made fairly explicit in the Franck Report, and their effect on the Acheson-Lilienthal Proposals was fairly obvious. At the time of the hydrogen bomb controversy the General Advisory Committee scientists did try, although without success, to separate their political and scientific judgments in their report. Today, however, the intensity of the intra-scientific conflict over nuclear weapons and the resultant increased politicalization of the scientist's advice have made it ever more difficult to know where scientific facts stop and political opinions begin.

Nevertheless, as the next chapter will show, the development of an increasing sense of social responsibility (and, therefore, political involvement) among scientists should not be regarded as cause for alarm. While it does exacerbate the problem of conflicting expertise, it can also constitute a source of great national strength. It is the task of political leadership to utilize more effectively than it has in the past the legitimate desire of scientists to contribute to the formulation of national policy toward nuclear weapons and other products of the scientific revolution.

CHAPTER TEN

THE TASK OF POLITICAL LEADERSHIP

In his engaging and controversial Godkin lectures[1] at Harvard University, Sir Charles Snow quite correctly characterized the problem of decision-making in the modern scientific world as "one of the most intractable that organized society has thrown up."[2] There are, Snow informs us, no "easy answers [otherwise] they would have been found by now." But there are, he continues, certain things which society must attempt to avoid in its decision-making, e.g., scientific overlords, scientific gadgeteers, and excessive secrecy.

Snow's major positive prescription is that there must be "scientists active in all levels of government" if society is not to stagnate but is to anticipate and meet the future effectively.[3] "Scientists," Snow affirms, "have it within them to know what a future-directed society feels like, for science itself, in its human aspects, is just that."[4]

As the present study has shown, American scientists, at least in the area of national policy toward nuclear weapons, have become full partners with politicians, administrators, and military officers in the formulation of policy. The American scientist has become a man of power to perhaps even a greater degree than Snow, his audience, or scientists themselves appreciate. Neither in any other nation of the world with the possible exception of contemporary Russia, nor in any other historical period, have scientists had an influence in political life comparable to that presently exercised by American scientists.

Yet, grave problems continue to exist with respect to the

[1] The lectures were published under the title *Science and Government,* Harvard University Press, 1961.
[2] *ibid.*, p. 67.
[3] *ibid.*, p. 80.
[4] *ibid.*, pp. 82–83.

government's utilization of scientists. Of these problems, the most fundamental throughout the history of the participation of scientists in the formulation of national policy toward nuclear weapons, and of policy in other areas as well, has been the failure of political leadership and of scientists themselves to appreciate fully the intricate relationship between the political and technical realms. Indeed, political leadership has tended to utilize scientists as if a clear distinction could be drawn between these two realms in the formulation of policy, and scientists have acted in the same fashion.

There have been many serious consequences of this fundamental error. On some occasions, as in the hydrogen bomb controversy and at the Geneva Conference of Experts, scientists have been assigned apparently technical responsibilities which in reality were as political as they were technical. On other occasions, as in the formulation of the Baruch Plan and of the American position on a nuclear test ban, the advice of some scientists has been acted upon without adequate criticism by other scientists or by experts in non-technical fields. Having been placed in positions where they have had to render essentially political advice or perform political functions, the scientists have had to exercise their political and strategic judgment as best they could. At times the judgment of the scientists has turned out to be wise, as were the recommendations of the GAC in 1949 for increased conventional military power. At other times, as at the Geneva Conference of Experts, the performance of the scientists has left much to be desired.

The failure of the scientist to appreciate the political nature of his activity has had a detrimental effect upon the scientific community. The belief of an individual scientist in his own objectivity has often resulted in the conviction that another scientist who disagrees with him is distorting the facts. When the political stakes have been high and emotions have been sufficiently stirred, this tendency among scientists has led to *ad hominem* attacks. Such was the case at Oppenheimer's security trial. Similarly, other scientists such as Edward Teller have been charged with intellectual dishonesty by scientists who disagree with them. "Dr. Teller believes," Harrison Brown told a University of Minnesota audience in 1958, "that any such agreements [to ban nuclear tests] would work to our disadvantage because we could not be certain that the Soviet Union might not 'bootleg' tests. I challenge this view, and in doing so I do not stand alone in the scientific world. I believe Dr. Teller is *wilfully distorting* the realities of the situation."[5]

Unfortunately, Brown, like so many of his scientific colleagues, has carried over into the political realm an attitude pertinent only to the treatment of scientific questions, the belief that the facts permit only one correct answer to which all intellectually honest scientists will give assent. Instead, as this study has argued, the proper question to be asked with respect to the scientist's advice to the public or to the government is not "Is it true?" but, "Is it wise?" To the writer's knowledge, despite the widely varying interpretations certain scientists have given to particular scientific facts, no scientist has been guilty of altering the facts themselves. On the contrary, one discovers that the American scientist's professional commitment to the truth has universally overcome his personal disappointment with the nature of the truth.

However, when the decision-maker asks the scientist-advisor what is the significance of a particular factor, such as the decoupling theory, for the technical feasibility of a nuclear test ban, the scientist's answer does not come solely from the realm of science. For although Hans Bethe would agree with Edward Teller on the validity of the decoupling theory, it seems to him that anyone who takes the theory as a serious obstacle to effective control is "behaving like a lunatic," while to Teller the theory represents the death knell of any effective control system. Certainly neither position can be adjudged to be true or false; yet the decision-maker must decide which scientist provides the wisest advice on this particular issue.

[5] Harrison Brown, "The Future of the Nuclear Race," *The New Leader*, March 31, 1958, p. 6. (Italics mine.)

Unfortunately the history of American nuclear policy reveals that the advice of scientists upon which political leadership has acted has not always been wise. Equally regrettably, that which has been wise has not always been heeded by American political leadership. This dual failure has been a major contributing factor to what this study characterized in Chapter One as the crisis of American nuclear policy. In essence, this crisis may be described as the inability of American political leadership to integrate nuclear weapons effectively into its national security policy. The attitudes of three successive national administrations toward nuclear weapons have been ambiguous ones; at the same time that these administrations have based American military policy primarily on the strategic and tactical employment of nuclear weapons, each administration has committed itself to the elimination of these weapons and, perhaps by implication, to refraining from the use of these weapons except in retaliation.

This situation has created a policy vacuum which scientists have on occasion felt impelled to fill. In a sense, then, to the extent that American political leadership resolves its nuclear dilemma, the political influence of scientists in this particular area will be decreased. The resolution of this dilemma, however, as well as the formulation of many other policies in the atomic age, will require the continued and active participation of scientists. For this reason, this study now raises and seeks to answer three questions: (1) Does the history of the participation of scientists in the formulation of American nuclear policy provide insights into the nature of the scientist—in Aristotle's term—as a political animal? (2) What has the intra-scientific debate over nuclear weapons policy contributed to an understanding of the requirements for a sound policy toward nuclear weapons? (3) What lessons has this intra-scientific debate provided for the more effective utilization of scientists in the formulation of public policy with respect to nuclear weapons and other products of the scientific revolution?

Critique of the Scientist As a
Political Animal

Prior to discussion of what this history of the intra-scientific conflict over nuclear weapons reveals about the scientist as a participant in political life, certain qualifications must be made. In the first place it must be recalled that this is a study of the *effective* public opinion and actions of scientists with respect to nuclear weapons policy. The study has not dealt with a random sample of scientists, but it has purposefully singled out for observation those persons who have been active in political affairs. In general, these are persons whose concern over national security, disarmament, and nuclear war has made them active in national politics. There is therefore the possibility that the political traits displayed by these scientists stem largely from their intense concern over national and human survival.

A second qualification is that this study has been largely restricted to American nuclear policy. One must grant, therefore, that scientists might behave much differently with respect to issues that affect them less directly or that evoke a less emotional response. Nevertheless, the writer believes that many of the political traits displayed by scientists in this history characterize the behavior of scientists with respect to other public issues. Further studies are needed of scientists in other areas of public policy such as health, economic, and foreign policies.

The third qualification to the conclusions of this study is one that is applicable to all social science research, i.e., its conclusions are probabilistic or statistical. There are, in other words, deviations from the norm; certain politically important scientists have exhibited a different attitude toward political life from that displayed by the vast majority of scientists. Perhaps due to experience, education, or adaptability, these scientists have not shared in equal measure with their colleagues the disposition toward political life which tends to characterize the scientist as a political animal.

The distinguishing characteristic of the scientist as a participant in political life stems from a fundamental ambiguity in the scientist's psychological make-up. At the same time that he appreciates the need for him to participate in political life, the scientist is often repelled by the requirements for success in politics. The appeal to passion, the skillful political maneuver, and the risk of public disfavor are highly repulsive to most scientists. Yet their sense of social responsibility, their desire to accomplish certain political goals, and even the desire for public honor draw scientists into political life. In short, the scientist wants to be both in and out of politics at the same time.

This problem for the scientist as a political participant arises from the difference between the intellectual and the political enterprises—a difference which has been thoughtfully summarized by Lyman Bryson in terms of three elements: the nature of conflict, legitimacy of tactics, and strategic approach.[6] Bryson merely argued that this difference was one of degree; yet the difference is significant enough to make Bryson's observations relevant to this discussion. Whereas, for example, intellectual, or in this particular case, scientific disputes are conflicts of ideas, political disputes are essentially conflicts of personalities. Upon the outcome of political conflict rest the fortunes and careers of the parties involved. Thus it was that the defeat of the GAC and the victory of Teller, Strauss, and Lawrence on the H-bomb issue led to the political decline of the former and the political rise of the latter.

The difference between the two enterprises is well illustrated by the Tizard-Lindemann conflict over radar and strategic bombing discussed by C. P. Snow. In a very real sense, Snow's attack upon Lindemann is one directed against an apostate. Lindemann's heresy, in the eyes of Snow and others, was to

[6] Lyman Bryson, "Notes on a Theory of Advice," *Political Science Quarterly*, Vol. 66, No. 3, September 1951, pp. 321–39.

break ranks with fellow scientists, to ally himself with a politician, and to enter without reservation into a struggle for political power with fellow scientists. Specifically, rather than acquiesce to the majority position among the scientists on the Technical Sub-Committee with respect to the feasibility of radar, Lindemann sought to defeat them through his alliance with Winston Churchill. For this apostasy and not simply because he was wrong on radar, Lindemann earned the lasting enmity of many scientists including that of novelist-scientist C. P. Snow.

The difference between science and politics which the Tizard-Lindemann case illustrates was well summarized in 1948 by the politician with whom Lindemann had allied: "The Committee[7] worked in secret, and no statement was ever made of my association with the Government, whom I continued to criticize and attack with increasing severity in other parts of the field. It is often possible in England for experienced politicians to reconcile functions of this kind in the same way as the sharpest political differences are sometimes found not incompatible with personal friendships. Scientists are, however, a far more jealous society. In 1937 a considerable difference on the Technical Sub-Committee grew between them and Professor Lindemann. His colleagues resented the fact that he was in constant touch with me, and that I pressed his points on the main Committee, to which they considered Sir Henry Tizard should alone explain their collective view. Lindemann was, therefore, asked to retire."[8]

Lindemann aroused antagonisms because he refused to be bound by the decision of his colleagues concerning the truth regarding radar, attempted to defeat Tizard through his alliance with Churchill, and had Tizard removed from govern-

[7] The Committee of Imperial Defence on Air Defence Research on which Churchill served and to which the Technical Sub-Committee reported. Lindemann had been appointed to the Sub-Committee as Churchill's personal representative. In fact, Lindemann's appointment was a condition of Churchill's decision to join the parent body. Winston S. Churchill, *The Gathering Storm*, Bantam Books, 1961, p. 135.

[8] *ibid.*, pp. 137–38.

ment in 1940. Such a political maneuver was contrary to the whole scientific ethos which holds that the scientific method is superior to politics as a tool for settling disputes among men. Lindemann was apparently placing loyalty to particular "gadgets" (as Snow calls them) above loyalty to the traditions and unity of science; thus he alienated his fellow scientists.

Both the Tizard-Lindemann example and the intra-scientific conflict over nuclear weapons reveal another implication of Bryson's observation that political disputes are personal conflicts while scientific conflicts are between opposing opinions. There is a tendency among scientists to overemphasize the importance of ideas in the political process. The scientist often attributes to ideas alone a cogency which, in the formation of political decisions, must be shared by political forces, peculiar circumstances, and the persuasiveness of particular individuals. Thus it is that scientists tend to view the hydrogen bomb controversy essentially as one between the ideas of the GAC and those of the Berkeley scientists. While the intellectual conflict was an important aspect of this history, it must be placed within the larger array of political forces.

This overemphasis on the role of ideas in politics has two consequences for the politics of scientists. In the first place it accentuates the bitterness of the intra-scientific political struggle. Believing that ideas have a special potency to move nations, scientists often compete bitterly to ensure a hearing for their ideas at the highest levels of government. In the second place, scientists tend to believe that political conflict can be settled if men have a clear understanding of the facts and issues involved. Politics should emulate science where disputes are settled by explanation and not by persuasion. Whereas the politician often seeks to persuade an opponent through appeals to passion, the threat of force, and the use of force, the scientist tends to assume there is one truth to which all reasonable men will accede once its nature has been explained.[9]

---

[9] An invidious comparison of these methods has been made by a scientist, Leo Szilard, *The Voice of the Dolphins,* Simon and Schuster, 1961, pp. 25–26.

A further aspect of the difference between the nature of scientific and political disputes is that whereas to the politician personal rivalry, such as that between Tizard and Lindemann and that between certain American scientists, is an expected part of the game, to scientists such conflicts appear due to a flaw in the character of one or both protagonists. As a consequence, personal rivalries among scientists tend to result in sharp and rather bitter cleavages in the scientific community. Politicians, on the other hand—as Churchill points out—being responsible for the continued survival of society and being aware that today's rival is tomorrow's ally, seldom permit personal rivalries to become complete ruptures.[10]

The difference between the ways of politics and the ways of science creates a situation where the scientist in political life is often caught between the requirements for political effectiveness and those for the approbation of his fellow scientists.[11] The emotional appeals of Pauling against nuclear testing, the lobbying of the Berkeley group for a hydrogen bomb, and the tacit alliance of Teller with Strauss on weapons development have alienated much of the scientific community from the individual scientists involved. And although the choice for the scientist is not always between political effectiveness and approval of his scientific colleagues, the dilemma is sufficiently acute to have caused one scientist, Edward Teller, to write in review of Snow's *Science and Government:* "And the reader is made to believe that Lindemann was successful as a politician simply because he became a politician. This may be the inevitable fate of those involved in politics."[12]

[10] Curiously, the possibility that national harm has resulted from the intra-scientific conflict over nuclear weapons has led one politician to suggest that his Senate committee conciliate the opposing factions of scientists in much the same way that some scientists in the Pugwash Movement seek to resolve the conflict between the politicians of East and West. Stuart Symington, *Congressional Record,* September 21, 1961, pp. 19390–94.

[11] An excellent analysis of this distinction is contained in Paul Nitze, "The Role of the Learned Man in Government," *The Review of Politics,* Vol. 20, No. 3, July 1958, pp. 275–88.

[12] Edward Teller, "Review of *Science and Government,*" *The New Leader,* June 5, 1961, p. 17.

Another difference between the intellectual and the political ways of life suggested by Bryson is one of strategy. Whereas the ideal of the intellectual or scientist is to base political action on a prediction, the politician's mode of action is to "initiate a maneuver and invent most of his planning as he goes."[13] The scientist is much more confident than the politician in the ability of man's mind to perceive the course of history and to devise a means by which to control it, and therefore he tends to elevate an insight or an analysis to the position of a general principle of political action.

Whereas the politician sees politics as a complex and unending enterprise which is often not amenable to human control, the scientist often believes his methods can cut through the web of complexities to a final resolution of the issues. Thus, whereas Jerome Wiesner, President Kennedy's science advisor, believes that, given the complexity and pace of the times, only large measures seem appropriate to resolve the problem of nuclear weapons, the politician believes as did Churchill in 1939 that "when events are moving at such speed and in such tremendous mass as at this juncture, it is wise to take one step at a time."[14]

This difference in approach between politicians and scientists has been a constant source of friction between members of the two professions. To the politician the scientist often appears to be a fuzzy-minded idealist. On the other hand, to the scientist the politician's commitment to "realism" makes him unreceptive to new, imaginative solutions (such as those proposed by scientists) to man's problems and, as a result, "many of the scientists would like to carve out . . . [an] area where they can carry out what they consider to be in the national interest in the realm of science without political interference."[15]

The desire of many scientists to exercise political power

[13] Lyman Bryson, *op.cit.*, pp. 327–28.
[14] See Chapter One for Wiesner's statement. For Churchill's statement, see Winston Churchill, *op.cit.*, p. 324.
[15] Paul Nitze, *op.cit.*, p. 284.

without becoming subject to political control or heeding the ways of politics is a source of constant danger for scientists. These men often fail to exercise the degree of prudence which is requisite to success in politics. Thus in 1949, President Truman, given the domestic and international political situation, could not have followed the moral and political recommendations of the GAC on development of the hydrogen bomb. By asking too much of political leadership, the GAC scientists only succeeded in decreasing their political effectiveness.

Study of the content of the scientists' solutions to political problems reveals that the scientists often neglect important political considerations in order to make their solutions to political problems plausible, and thus in a real sense they do tend to provide essentially "technical" solutions to political problems.[16] This is to say that although their policy proposals rest on certain non-technical or political assumptions of a general and almost philosophical type, political factors as such too seldom enter into their analysis. To scientists, political reality does not always constitute a "given" which must be taken into account; instead, it is often assumed to be a malleable superficiality which man's reason can change in accordance with his heart's desire.

The failure of scientists to recognize and to be sufficiently concerned with the political aspects of problems to which they have proposed solutions is illustrated by the intra-scientific struggle over nuclear weapons. Thus, for the pro-ban scientists, the task of American disarmament policy has been to devise some technique, method, or system which will make disarmament possible. Assuming, as did the Franck Report, that "lack of mutual trust, and not lack of desire for agreement" is the major obstacle to efficient disarmament, these scientists believe

---

[16] For a discussion of "technical" solutions see Charles J. Hitch, "Character of Research and Development in a Competitive Economy," a paper appearing in Proceedings of a Conference on Research and Development and Impact on the Economy, National Science Foundation 58–36, 1958, pp. 129–40.

that the real challenge has been and is to devise a technical means to implement an implicit East-West political agreement upon ends.

On the contrary, the crux of the disarmament problem has primarily been the absence of the desire for agreement and not simply the absence of mutual trust or of technical means. The United States, viewing massive retaliation with nuclear weapons as its principal deterrent to Communist aggression, had, at least until the arrival of mutual deterrence, little incentive to take arms control seriously. Regrettably, even though the United States has within the past few years looked at arms control in positive terms, the Soviet Union not only continues to show little genuine interest in it but indeed now views nuclear blackmail as a valuable instrument of national policy.

If the nuclear test ban negotiations have proven anything, it is that the principal political obstacle to the achievement of agreement has been a lack of will and not of technology. While technical advance may contribute to the political feasibility of a project, the technology does exist which can be used to police a test ban, *provided* that the Russians become willing to make a few political concessions such as permitting on-site inspection of Russian territory, the establishment of manned and robot detection stations in Russia, and agreement upon a workable control commission.

In the light of this interpretation, the task for American policy is to increase the incentive of the Soviet Union to negotiate arms control with sincerity. Only when the Soviet Union learns that it has more to gain through arms control than it can attain through a policy of nuclear blackmail can there be much hope for the success of arms control negotiations. "As the free world gains in purpose, cohesion and safety," Henry Kissinger has written, "the Communist approach to negotiations may alter. Instead of using arms control negotiations to tempt or blackmail the West into unilateral disarmament, the Communist leaders may address themselves seriously to the problem of how to reduce the tensions inherent in an unchecked arms race. Then co-existence may become something other than a slogan."[17]

Turning to a consideration of the arguments of the anti-ban scientists for a Western strategy which places principal reliance upon a limited nuclear war and rejects measures to control the nuclear arms race, the same type of political myopia which characterizes the pro-ban scientists is found. A noteworthy example is their evaluation of nuclear and conventional weapons solely in technical terms and their belief that the only difference between the two is the difference between an efficient and an inefficient weapon. Instead, it must be realized that war and the means for its making are instruments of policy and must be evaluated from this perspective. A weapon whose employment may create greater political liabilities than military assets is not the instrument upon which all of a nation's political commitments should be based.[18]

When the anti-ban or infinite containment scientists oppose arms control because there is a danger of a "competitive premium on infringement" in a partially closed world, they may again be myopic. This position ignores the real possibility that the purpose of an arms control measure may be solely political and that the political goal might outweigh military risk. Thus a cogent argument for a nuclear test ban, including a ban with only modest inspection provisions, is that it would symbolize a tacit agreement among the nuclear powers not to employ nuclear weapons in limited war. Such a ban would then be likely both to discourage the have-not powers from investing their scarce resources in nuclear weapons and to inhibit the nuclear powers from utilizing these weapons in limited war.[19] As a matter of fact the test ban negotiations

[17] Henry Kissinger, *The Necessity for Choice*, Harpers, 1961, p. 7.

[18] *ibid.*, Chapter Three is an excellent critique of the limited war challenge facing the United States. For an additional penetrating critique see Morton Halperin, "Nuclear Weapons and Limited War," *Journal of Conflict Resolution*, Vol. 5, No. 2, June 1961, pp. 146–66.

[19] See Donald G. Brennan and Morton Halperin, "Policy Considerations of a Nuclear-Test Ban" in Donald G. Brennan, ed., *Arms Control, Disarmament, and National Security*, George Braziller, 1961, Chapter Twelve.

themselves may have served this purpose. However, whether this argument has validity is not at issue here; the point is that the arguments of the anti-ban scientists fail to recognize the possible political effects of arms control and of nuclear weapons.

Both sets of scientists are really in basic agreement on fundamentals. Both sides reject the concept of political life held by political theorists and politicians—that of a never ending struggle for power among men and nations with opposed interests. Both sets of scientists assume that the end of political conflict is achievable due to the existence of nuclear weapons; whereas one side (finite containment) believes that the need to cooperate in controlling nuclear weapons gives nations a mutual interest which overrides all their other conflicts, the other side (infinite containment) argues that the threat to employ nuclear weapons can eliminate the will of nations to war one upon another.

Both groups of scientists reject the middle view that opposed national interests will continue to be sources of conflict and war despite the fact that the danger of nuclear war does give nations a new mutual interest. The finite containment scientists tend to disregard these areas of interest conflict in their elaboration of arms control measures; similarly, the infinite containment scientists, when they reject arms control, fail to appreciate the fact that these conflicts of interest might lead to a full scale nuclear war which no one wants and which might be prevented through appropriate arms control measures. In short, neither side fully recognizes that arms control measures are required because of the present world political situation and that they themselves require for their successful implementation a full appreciation of the struggle for power among the nations.

Due to their strong and understandable desire to solve the security problem facing the United States and, of course, all humanity, scientists have too often permitted dispassionate and rigorous analysis to give way to nostrums, oversimplification, and preconceived solutions. Their approach to politics has frequently been similar to that which E. H. Carr described in the 1930's when many persons were making various proposals to prevent the fast approaching clash of arms; the political thinking of too many scientists is still largely at that stage of development when "investigators will pay little attention to existing 'facts' or to the analysis of cause and effect, but will devote themselves whole-heartedly to the elaboration of visionary projects for the attainment of the ends which they have in view—projects whose simplicity and perfection give them an easy and universal appeal." Instead "the end [sought by many scientists] has seemed so important that analytical criticism of the means proposed has too often been branded as destructive and unhelpful"; the scientist has too often "replied to the critic not by an argument designed to shew how and why he thought his plan will work, but either by a statement that it must be made to work because the consequence of its failure to work would be so disastrous, or by a demand for some alternative nostrum."[20]

While recognizing the special attitudes and problems of scientists in government one should also inquire into the manner and degree to which scientists have adapted to political life. Indeed, has the participation of scientists in political life since World War II modified their political attitudes and made them more appreciative of the ways of politics? Such a development is dependent upon the scientists' willingness to see politics in a new light, to appreciate the irrationalism of political life, and to gain a sympathy for the problems faced by politicians.

In the words of one observer, Paul Nitze, who has served in government with many scientists, " . . . a special effort is required for the expert to abandon concentration upon his chosen field and broaden his interests and temper his judgments to that humanity which is necessary to deal success-

[20] E. H. Carr, *The Twenty Years' Crisis, 1919–1939,* London: Macmillan and Co., Inc., 1951, pp. 5, 8.

fully with politics in a democracy."[21] In an effort to move in this direction, Vannevar Bush has admonished scientists to respect "those individuals who are masters of the art of operating in the confused area of the American political scene . . . . In fact," Bush continued, "if scientists are to have their full influence for the good of the country in the days to come, many of them will indeed need to learn to practice this difficult art."[22]

Nitze's and Bush's comments remain pertinent; nevertheless, it is obvious that many scientists are re-educating themselves, broadening their interests, and tempering their judgments. Since 1945, the tone of the political writings of many scientists has developed political sophistication. Recent literature, for example, reveals scientists to be more modest, analytical, and appreciative of the intractable nature of political problems than they were in 1945. In one instance a scientist has written a major political work; whether or not one agrees with the analysis in Herman Kahn's *On Thermonuclear War*,[23] it is a work which should command the respect and attention of anyone who wants to understand the problem of modern warfare.

Furthermore, it is important that the scientist's immense contributions to American political life should be acknowledged. Motivated by their concern over a "solution" to the overall problem created for America in the nuclear age as well as by a sense of duty to country, American scientists have selflessly devoted themselves to the common good through active participation in political life. This participation has resulted in policy initiatives, organizational reforms, and weapons innovations.

[21] Paul Nitze, *op.cit.*, p. 285.

[22] "The Scientist's Role in Political Decision Making." Statement for 15th National Conference on the Administration of Research, October 10–13, 1961, San Juan, Puerto Rico.

[23] Princeton University Press, 1961. For other examples of political analysis by natural scientists see the essays in Donald G. Brennan, ed., *Arms Control, Disarmament, and National Security, op.cit.;* David Frisch, ed., *Arms Reduction-Program and Issues,* The Twentieth Century Fund, 1961.

The Contribution of Scientists to a
Sound Nuclear Policy[24]

Both the finite and infinite containment schools have made valuable contributions to our understanding of the problems and opportunities presented by nuclear weapons. Certain ideas of each school seem to have withstood the test of experience of the past fifteen years and of criticism by opposed points of view. For these reasons, it would appear prudent that a sound nuclear policy take the contributions of each group of scientists into account.

CONTRIBUTIONS OF THE FINITE CONTAINMENT SCIENTISTS

The first contribution of the finite containment scientists is their contention that nuclear weapons have brought the world into a new and exceedingly dangerous era whose outlines man can as yet only dimly perceive. The attempts of these scientists to grasp the realities of this new world have contributed to our understanding and have also stimulated insights by social scientists into the novel aspects of international politics in the atomic age.[25] These natural scientists have made three points which help to define the modern age.

[24] As the activities of the control school (in this writer's opinion) have been solely polemical and not analytical, the views of these scientists are not included in this summary. While the scientists in this school have aroused anxiety about the dangers of nuclear weapons and their testing and have had an indeterminate political influence, they have made few contributions to our understanding of the problems raised by the advent of nuclear weapons.

[25] The issue of how novel international politics is in the atomic age is one that divides many thoughtful persons. Most natural scientists and some social scientists believe that atomic weapons have produced a discontinuity in history and that the very nature of international relations has been changed. See, for example, John Herz, *International Politics in the Atomic Age,* Columbia University Press, 1959.

Most students of international politics appear to assume that although the impact of atomic weapons on international politics has been very great and has changed many of its aspects such as the nature of war, it has not changed the fundamental nature of international politics. The validity of this approach is attested by the fact that writers in 1945 who made an assumption of historical continuity have thus far been relatively more correct than those who did not make this assumption in their predictions for the atomic age. See, for example, Jacob Viner, "The Implica-

In the first place, their notion that nuclear weapons have outdated the conviction that political settlements must precede or at least be undertaken in conjunction with disarmament or arms control appears to have some validity. While disarmament and arms control are not the autonomous and solely technical problems which these scientists tend to assume, some credible arguments have recently been made by political analysts within and without the scientific community that some progress might be made through treatment of arms control measures as a semi-autonomous area for international negotiations.[26]

The second valuable idea contributed by the finite containment scientists is closely related to the first. It is that nuclear weapons are a cause as well as a symptom of international political tensions. The magnitude of the destructive power of these weapons has created severe political problems with respect to their spread, deployment, and threatened use. Although it may well be argued that these scientists tend therefore to overestimate the possible beneficial political effects of measures to control nuclear weapons, there can be little doubt that some control measures which would increase world stability would be politically beneficial.

Yet another tenet of these scientists which has political relevance is their belief that the United States and the USSR, despite their grave mutual antagonisms, have a strong mutual interest in taking steps to prevent nuclear warfare. Although they undoubtedly overemphasize the extent of this mutual interest in arms control, they quite correctly do stress this

---

tions of the Atomic Bomb for International Relations" in Harlow Shapley, ed., *A Treasury of Science*, Harpers, 1946, pp. 751–60.

An excellent analysis of the impact of nuclear weapons on international politics is Kissinger, *op.cit.*, Chapters Two and Three.

[26] The literature on this subject has grown impressively in the past two years. In addition to the works already cited by Brennan, Frisch, and Kahn, there is the excellent study by Thomas C. Schelling and Morton H. Halperin assisted by Donald G. Brennan, *Strategy and Arms Control,* The Twentieth Century Fund, 1961.

factor as an important consideration in policy-making. In particular, they emphasize the need for the United States to develop policies which will encourage the Soviet Union to modify its policies tending toward accidental war. Even though these scientists do sometimes have rather questionable notions of the sources of Russian behavior and overemphasize the degree to which the United States could modify Soviet behavior, they are undoubtedly correct in believing that the United States could influence the direction of Soviet policy in a beneficial direction far more than it has in the past.

On the other hand, although the finite containment scientists have proposed ideas which should be incorporated in the formation of United States nuclear weapons policy, these men have failed to recognize certain other repugnant features of the atomic age which must be part of one's analysis of the problem of nuclear disarmament and arms control. They neglect, for example, to appreciate the implications for arms control of the political value to aggressive powers of the phenomenon of nuclear blackmail, e.g., the threat to wage nuclear warfare if political demands are not met. The threat of nuclear blackmail, as the world has observed in the Berlin crisis, has become an established tactic of Russian policy. The Russian resumption of nuclear tests in the fall of 1961 and the threat to destroy other nations with a nuclear weapon equivalent to 100 million tons of TNT were in part an attempt to play on the world's fears of nuclear weapons in order to gain political ends. It is significant that the Russians have made over 125 nuclear blackmail threats against the Free World in recent years.[27]

Despite the tendency among members of the finite containment school to avoid facing certain of the harsh realities of the atomic era, these scientists have emphasized a number of specific problems with which American nuclear policy must

[27] For a list of some of these nuclear blackmail ventures see U.S. House of Representatives, Committee on Foreign Affairs, *Background Information on the Soviet Union in International Relations*, 87th Congress, 1st Session, 1961, pp. 41–50.

be concerned. The first of these is the potential spread of nuclear weapons to more and more nations; the second is the difficulty of keeping any limited nuclear war limited; and the third is the need for the United States to increase its capabilities for limited conventional warfare.

These concerns of the finite containment scientists have led them to reject the thesis that American superiority in tactical nuclear weapons can deter limited war and win it if deterrence should fail. Instead, their belief is one first stated by Robert Oppenheimer in the early days of the atomic age and amplified in Project Vista; it is that the United States must not become overdependent upon nuclear weapons. Unfortunately, both the Truman and Eisenhower Administrations tried to obtain security almost solely through dependence upon nuclear weapons which they, for political and strategic reasons, dared not use. The position of the West in the early 1960's with regard to the problems of Berlin, Laos, and South Vietnam reveals how correct these scientists have been in their insistence that the West needs to possess greater conventional military power in order to meet its political commitments.

In summary, then, these scientists with a strong desire to see nuclear weapons eliminated or at least brought under some system of arms control, have made a number of positive contributions to an understanding of the problems facing the United States. Their commitment to the international control of nuclear weapons, however, has blinded them to other aspects of the atomic age which confound the achievement of their political goals. For a recognition of some of these other matters one must turn in part to the thinking of their scientist-opponents in the infinite containment school.

CONTRIBUTIONS OF THE INFINITE
CONTAINMENT SCIENTISTS

The scientists in the infinite containment school contribute an important consideration necessary for a sound under-standing of the problems and—as they would argue—the opportunities presented by nuclear weapons. Their point that the combination of secrecy and rapid technological advance could quickly undermine a system to "control" armaments is their most significant contribution. The Soviet atomic bomb, the Soviet hydrogen bomb, and the technical undermining of the Geneva System appeared much faster than even the most pessimistic American scientists had anticipated. The infinite containment scientists therefore believe that any sound nuclear policy, especially in the area of arms control and disarmament, must assume equally rapid advances in the future. As any measures to arrest technological advances place a "competitive premium on infringement," these scientists repeatedly emphasize that international agreements such as a *total* ban on nuclear weapons testing would constitute an unacceptable risk save in an open world.

Whether or not the above argument of the infinite containment scientists is fatal to the possibility of an inspectable test ban can certainly not be settled here since information necessary to reaching a decision is classified. However, one can point out that the existence of secrecy and rapid technological advance make rather futile most suggestions proposed thus far to end the nuclear arms race. Unfortunately, the finite containment scientists are slow to concede that an "open world" and world government are actually prerequisites for many of the arms control and disarmament measures they propose. It was, however, a strong proponent of such measures and not a member of the infinite containment school who wrote that: "any *serious attempt at inspecting research and development activities implies the complete elimination of secrecy from research.*" For this reason, the writer continued, in order to achieve the type of technical stability essential to the success of any significant disarmament system, there must be "the eventual relinquishment of absolute national sovereignty in favor of an international order . . . ."[28]

[28] Bernard Feld, "Inspection Techniques for Arms Control," *Daedalus* (*Arms Control* Issue), Vol. 89, No. 4, Fall 1960, p. 877. (Italics mine.)

A second contribution of the infinite containment scientists to an understanding of the requirements of a sound nuclear policy results from their constant emphasis on the strategic and political importance of nuclear weapons. They stress that although the destructiveness of these weapons may in one sense make their elimination desirable, it also gives these weapons a decisiveness and political significance which appear to make their elimination impossible.

These scientists remind us that it is technically impossible within present levels of knowledge to devise any methods to ensure compliance with any international agreement to eliminate stockpiles of nuclear weapons. It will be remembered that in 1955 the United States, on the basis of just such reasoning, decided the goal of American nuclear policy ought to be peace through mutual deterrence rather than the complete elimination of nuclear weapons.[29]

An implication of the destructiveness of nuclear weapons emphasized by the infinite containment scientists is the point first made in 1945 by the scientific advisory panel to the Interim Committee that the threat posed by nuclear weapons can be a deterrent to a major war between East and West. That these weapons (assuming they are not abandoned by East and West) shall continue to fulfill this deterrent function is the challenge these scientists pose for a sound nuclear policy.

In relation to the Nth country problem, these scientists force us to realize that a realistic approach to this serious subject requires a more profound analysis of why nations would seek to acquire nuclear weapons in the first place than that implicit in the proposal for a nuclear test ban. For example, some nations such as Communist China[30] probably

---

[29] The Russians concurred in their May 10, 1955 proposal but drew different political conclusions.

[30] Many advocates of a nuclear test ban engage in an intellectual sleight of hand on this point. They argue the feasibility of a test ban as a first step in terms of Russia's interest in keeping Communist China denuclearized. However, they do not apply their *realpolitik* analysis to explain

cannot be denied nuclear weapons short of a major war waged by the United States and the USSR in alliance against China, a highly unlikely development. Other nations such as the members of NATO might be deterred from gaining independent nuclear capabilities through some system of nuclear sharing wherein the United States would continue to exercise control over the weapons. Lastly, a nuclear test ban for a variety of reasons might deter some other nations from investing their scarce resources in nuclear weapons.

One can agree, however, with the basic notion of these scientists that nuclear weapons can and have thus far served as a deterrent to major war and, to a lesser extent, to limited war between the nuclear powers without accepting the related prescription that the West ought to depend primarily upon tactical nuclear weapons to meet limited Communist aggression. The problems of the escalation of limited nuclear war into total nuclear war and the political consequences which would be incurred by the use of such weapons provide serious disadvantages to a policy which bases resistance to piecemeal Communist aggression principally upon these weapons.

THE CHALLENGE TO AMERICAN NUCLEAR POLICY

Even though the contributions of each set of scientists to an understanding of the problems and, as some would argue, the opportunities presented by nuclear weapons may be accepted as valid, yet it is difficult to fashion these ideas into a sound policy toward nuclear weapons. There can be little doubt, for example, that it is impossible to reconcile the argument of certain scientists for the total elimination of nuclear weapons with the argument of others that, save in an open world, such an accomplishment would only place a "competitive premium on infringement" of the

---

why China would have an interest in denying itself nuclear weapons. A test ban openly defied by China would be very demoralizing for any long-range hope to solve the problem of nuclear weapons.

control system. For this reason, if one accepts as genuine the commitment of the United States since September 25, 1961, to seek universal and complete disarmament, he must do so in the face of cogent arguments against the wisdom of such a policy except perhaps as a propaganda move in the Cold War.

On the other hand, if one believes that the real goal of American nuclear policy should continue to be what it was from August 1955 to September 1961, namely, the control rather than the elimination of nuclear weapons, then it is not inconceivable that these opposed positions of the scientists can be reconciled.

American political leadership must resolve what this study referred to in Chapter One as the contradiction in American nuclear policy. This is the conflict between American disarmament policy which seeks to prevent the usage of nuclear weapons and if possible to eliminate them completely and American military strategy which seeks to defend American political commitments by means of the threat and, if necessary, the use of nuclear weapons.

A major and dangerous consequence of the continuation of this contradiction has been to make less credible the American threat that it will employ nuclear weapons to defend its political commitments such as those with respect to Berlin. Until the West is capable of defending such commitments by conventional arms, it would appear highly imprudent for the United States to propose disarmament measures which detract from the credibility of its nuclear deterrent. To do so would be to risk nuclear war through Russian miscalculation of Western determination. Yet, to ignore the long-range threat of nuclear weapons to mankind is to court disaster. Somehow, and in some way, the United States must resolve the basic contradiction in its nuclear policies. It must pursue arms control measures which will reduce the possibility of nuclear war at the same time that it maintains the credibility of its nuclear threat. The United

States capability and will to use nuclear weapons must be maintained until such time as the West can meet its political commitments by other means. Fortunately, under President Kennedy's leadership, an attempt is being made to achieve this objective.

A solution to the contradiction in American nuclear policy requires a more modest approach to the problem of arms control and disarmament than those which have been undertaken since 1945. Commencing with the Baruch Plan and continuing through the Open Skies and nuclear test ban proposals, the United States by trying to accomplish too much has accomplished too little. Even a nuclear test ban requires rather far reaching concessions from both sides and, under President Kennedy's disarmament proposal of September 25, 1961, the test ban continues to be the necessary first step to any arms control agreements between East and West. As a consequence of this emphasis on a nuclear test ban as a first step, lesser measures may be neglected which would probably help to stabilize the world.

There is no greater problem facing the United States today than that of simultaneously defending its political commitments while seeking means to lessen the probability of thermonuclear war. The history of the intra-scientific conflict over nuclear weapons illustrates the difficulty of the task and the contributions of each set of scientists to an understanding of the problem. It is hoped that this history has indicated the need for a more effective utilization of scientists as the problems of the atomic age are confronted.

Toward a More Effective Utilization of Scientists

This study has pointed out a number of weaknesses in the government's utilization of scientists. Firstly, American political leadership has failed to come to grips with the problem of conflicting expertise and to appreciate its significance for the attainment and utilization of wise advice from scientists. Secondly, successive administrations have failed to integrate the scientists' advice on nuclear weapons with political, military, and other considerations; instead the advice of scientists on occasion has played too predominant a part in directing the formulation of American nuclear weapons policy and, at other times, it has been ignored despite its valid criticisms of the trend of American policy. Lastly, the activities of the scientist as advisor have raised certain issues of the scientist's rights and responsibilities. For all these reasons political leadership must adopt measures to make more effective the scientist's active participation in the formulation of national policy toward nuclear weapons and toward other results of scientific advance.

PROPOSED SOLUTIONS TO THE PROBLEM OF CONFLICTING EXPERTISE

When this study discussed the problem of conflicting expertise, it placed in juxtaposition two apparently conflicting statements by Edward Teller and Hans Bethe on the technical feasibility of a nuclear test ban. The discerning reader may have observed that the conflict between these statements is dependent upon the manner in which various terms in each statement are interpreted. What did Bethe mean by "technically feasible," "reasonable assurance," and "very small, decoupled tests?" Or what did Teller mean by "knowing," "technical methods," or "to police?" As has been shown in Chapter Nine, a number of elements in the debate such as the qualitative nature of the terms and the employment by each side of *ad hoc* hypotheses prevent agreement upon definition of terms and thwart a resolution of the issue in conflict. Still, it is widely recognized that the problem of conflicting expertise must be solved if political decision-makers are to be able to evaluate and act wisely upon scientific advice. Worthy of discussion are three major proposals which have been put forth as solutions to the problem; each has some merit despite serious limitations of applicability.

One proposed solution is that the scientist-advisor ought to be educated to the necessity of maintaining a strict separation between his technical knowledge and his political opinion. In essence, the notion that this solution is feasible underlay the prosecution's allegations at the time of Robert Oppenheimer's security trial: his strategic and political views had influenced his advice on the advisability of a crash program to develop the hydrogen bomb. Throughout the trial the implication was present that the scientist has a responsibility to advise only on the "technical" facts and that he has no right to include political or other views in his advice to decision-makers.

This position that the scientist must be educated to maintain a strict isolation between his scientific and political ideas rests on a simplistic notion to which this discussion takes exception, the idea that the technical and political realms can be kept separate in the scientist's advice. Indeed they cannot; it is not as if a Teller or a Bethe were deliberately inserting political opinion into his views on the technical feasibility of a nuclear test ban. Each believes that his position is the scientifically correct and objective one; each believes that it is his opponent who has interjected political considerations into his advice.

The problem of scientist-layman communication further complicates the difficulties in distinguishing technical knowledge from political views. By necessity the scientist-advisor must select from all available scientific facts those which he believes are relevant to the problem under consideration. Furthermore, if these selected facts are to be meaningful to political leadership, the scientist must give them meaning. In this selective and interpretive process the scientist's own political views become an integral part of his advice to political leadership.

The writer believes that therefore the suggestion that the solution to the problem of the political nature of the scientist's advice lies with the scientist himself is quite unrealistic. While true objectivity should continue to be a goal of the scientist-advisor, the frequent intertwining of the technical and political aspects of the scientist's advice make this a desirable goal which is seldom attainable.

A second solution which has been proposed for this problem of communication between the scientific expert and the political leader is the notion, often expressed by scientists themselves, that political leaders ought to be educated sufficiently in science to be able to ask meaningful, strictly technical questions of scientists. Scientists rightfully complain that, throughout the postwar period, political leaders have asked them questions which go beyond their competence and at times have not tried to understand their answers.

Eugene Rabinowitch has spoken for many scientists on the need to educate political leadership in science as a solution to this problem. Rabinowitch writes in relation to the conflict among experts on the harmful effect of fallout: " . . . if the questions put to scientists were adequately formulated and the answers received properly understood, the public and the political leadership would have easily found out that the two groups of experts [on radioactive fallout] had nearly the same answers . . . . The important thing is how to ask scientists questions which science can answer, to appreciate the authority with which science can speak under certain circumstances and its limitations in others."[31]

However, it must be realized that this ability to ask the right questions may require a scientific competence which one acquires only through extensive scientific training. Thus, with respect to the matter of the technical feasibility of a nuclear test ban, it required a highly competent scientist to

---

[31] Eugene Rabinowitch, "Decision-Making in the Scientific Age," *The Scientific Revolution,* Edited by Gerald W. Elbers and Paul Duncan, The Public Affairs Press, 1959, pp. 24–25.

inquire of his colleagues whether or not the detonation of nuclear explosions in large underground cavities would make them difficult to detect. Nevertheless, in the present age, an understanding of science and an ability to frame intelligent questions for scientists ought to be encouraged among public officials. Fortunately, many politicians have become proficient in this respect.

A third suggestion for the solution to the problem of conflicting expertise has been advocated by James Conant. Conant has written that because "it is so easy for the proponent of a project to clothe his convictions in technical language," the government's advisory mechanisms must find "ways and means of balancing the biases of experts whenever their opinions are of prime importance in the making of decisions."[32]

The most important step advocated by Conant is the employment on advisory committees of experts who have opposed emotional biases. If this is not possible, then some experts must be assigned the task of devil's advocate.[33] Insofar as it is practicable, advisory bodies ought to be consciously composed so as to include scientists with opposed political points of view. The effect of the direct confrontation of scientists with opposed political assumptions and values is to raise to a conscious level the implicit non-technical assumptions which frequently cause apparent scientific disagreement while leaving intact the points of scientific agreement. Confrontation makes the political ideas of the scientists explicit and accessible for evaluation by decision-makers.

This solution to the problem of conflicting expertise also provides one major guide to the selection of scientists as advisors. The task of political leadership is not, as C. P. Snow seems to suggest, to seek out the "good" scientists (such as a Sir Henry Tizard) as advisors and to avoid

---

[32] James Conant, *Modern Science and Modern Man,* Doubleday, 1952, pp. 114–15.
[33] *ibid.*, pp. 115–18.

reliance upon the "bad" scientists (such as an F. A. Lindemann). The history of the participation of American scientists in the formulation of nuclear policy reveals that the advice of no one scientist or group of scientists has always been infallible and wise. The history does indicate, however, that wisdom flourishes best and error is avoided most effectively in an atmosphere of intellectual give and take where scientists of opposed political persuasions are pitted against one another.

In the period since 1945, with the possible exception of the years from 1954 to 1958, confrontation of various scientific points of view has been made possible in government by the diversity, complexity, and competitive nature of the American political system. Although one scientific point of view or another may have dominated the principal scientific advisory bodies concerned with a particular issue, scientists with opposed views have been able to make themselves heard through contacts with other influential groups within the government. In 1949, for example, the infinite containment scientists, though not on the General Advisory Committee, were able to make known to the President their views on thermonuclear weapons through discussions with such persons as AEC Commissioner Lewis Strauss, Chairman Brien McMahon of the Joint Committee on Atomic Energy, and the minority leader of the Senate, William Knowland.

The history of President Eisenhower's decision to seek a nuclear test ban, regardless of the merits of the decision itself, reveals the value to the President of many sources of scientific advice. The formation of the President's Science Advisory Committee in 1957 made available to the President a source for advice which contrasted to the negative advice he had received from the AEC and the Pentagon on the wisdom and feasibility of a nuclear test ban. As a consequence the horizons of the President were expanded and his range of choice was increased.

The recommendations of scientists, as Robert Oppenheimer noted with respect to the GAC's recommendations on the hydrogen bomb, tend to be "total views where you try to take into account how good the thing is, what the enemy is likely to do, what you can do with it, what the competition is, and the extent to which this is an inevitable step anyway."[34] A most effective way to separate such "total views" into their technical and political components is, as Conant suggests, the confrontation of opposed points of view.

This fact plus the experience of President Eisenhower in the nuclear test ban issue illustrate the wisdom of Don Price's observation that "since many important issues can be decided only on the basis of organized study, and since for many reasons it is impossible to have competing studies, the right to conduct the study or prepare the plan becomes almost equivalent to the final authority to decide. The executive's only opportunity for control is then the very general and long-range one of controlling the nature of the advisory or planning machinery and the selection of its top personnel."[35]

MEASURES TO IMPROVE THE INTEGRATION OF TECHNICAL AND NON-TECHNICAL ADVICE

American scientists have also made two valuable contributions to the process of national policy making. The first is the notion that the scientist, operating at the frontiers of knowledge, is an anticipator of political and social problems posed by scientific advance; he is, as C. P. Snow has put it, "future-directed" and is sensitive to the difficulties scientific advance will cause for mankind. The second contribution of the scientist to national policy-making, although certainly not his alone, is the notion that the government

---

[34] U.S. Atomic Energy Commission, *In the Matter of J. Robert Oppenheimer*, U.S.G.P.O., 1954, p. 80.

[35] Don K. Price, *Government and Science*, New York University Press, 1954, p. 158.

ought to apply scientific analysis to the solution of its problems such as those generated by scientific advance. Summer study projects, the panels of PSAC, and research contracts to private institutions all reflect this innovation in the process of policy making for which the nation is indebted principally to the natural scientists.

However, as this study has shown, there are two necessary qualifications that must be made to these points. In the first place, although the scientist has proved in many instances to have been prescient concerning technological development, he has just as often been wrong. In 1948, just one year prior to the first Soviet atomic explosion, Robert Oppenheimer and other scientist-advisors were informing the Administration that the Russians would not have an atomic bomb for a long time to come. Scientific advisors displayed similar optimism regarding Russian development in the thermonuclear and missile fields.

In the second place, the scientist has not always been correct in his assessment of the social and political implications of technological developments. Furthermore, even in those cases where the scientist has been correct in anticipating the problems scientific advance causes for society, he has not been very successful in proposing solutions to these problems. This is undoubtedly because the natural scientist has lacked a sufficient appreciation of the social world and its peculiar dynamics. Thus, even though the world in the second decade of the atomic age is characterized by many elements predicted by scientists in 1945, major factors in the international scene such as the existence of mutual deterrence, the Soviet policy of nuclear blackmail, and the re-emergence of limited war as a major element of national strategy, were not part of the scientists' prophecies. Moreover, the solutions to the nuclear arms race which were put forth in the Bohr Memorandum and the Franck Report have not proven efficacious.

Similarly, although the scientist has made a valuable contribution to decision-making in his emphasis on the notion that the government ought to apply scientific analysis in the formulation of its policies, in doing so he has too often brought into his own analyses of political problems attitudes solely appropriate to the study of natural phenomena. As a consequence, the natural scientist paradoxically has tended both to expect far too much from a scientific approach to policy-making and to be unaware of the fact that political matters are researchable; the scientist in politics too often searches for *the* solution to a problem and gives insufficient attention to the political and strategic aspects of either the problem or its solution.

Unfortunately these tendencies of the scientist are complemented by predispositions frequently held by the policy-maker: (1) "the assumption that the solution of any problem will be advanced by the simple collection of fact;"[36] (2) a tendency of harassed and sometimes indecisive policy-makers to accept uncritically expert advice confidentially given; (3) a predilection of the politician to trust the technical specialist over the administrative generalist; and (4) an understandable propensity among governmental officials to permit the requirements of immediate action to supplant long-range reflection.[37]

There exists, then, a need to develop means by which the natural inclination of the scientist to advocate technical solutions to political problems and the tendency of the administrator to accept uncritically such solutions can be countered. Effective ways of integrating the technical and political knowledge in policy-making must be found. Only in this way can the United States avoid further errors like those which this history of United States nuclear policy has revealed.

[36] Max F. Millikan, "Inquiry and Policy: The Relation of Knowledge to Action," in Daniel Lerner, ed., *The Human Meaning of the Social Sciences,* Meridian Books, Inc., 1960, p. 163.
[37] See Henry Kissinger, *op.cit.,* pp. 340–48.

One technique to bring about the integration of the technical and political aspects of policy has been innovated by the scientists themselves. This is the contracted study project, such as Project Vista, wherein experts from a variety of fields and from both inside and outside of the government meet together over a period of months to fashion policy suggestions in a broad area of national concern. The results of past projects indicate that such a procedure can create ambitious and significant proposals and can also avoid the frequent tendency of the interdepartmental committee to agree only on the lowest common denominator among differing positions. It is hoped that such projects will be one consequence of the recently established United States Arms Control and Disarmament Agency whose task it is to develop American arms control and disarmament policies.

A second way in which technical and political aspects of national policy can be integrated is the appointment of scientific advisors to advise the major responsible officials in the government. Increasingly, science is permeating the activities of almost every government agency and is being affected by the policies of officials whose primary concern is not science itself. Thus, while the existence of the President's Special Assistant for Science and Technology and the Science Advisory Committee has done much to bring science into national policy-making, the potential contribution of science to national welfare will be realized only when scientists are active in all federal agencies. Furthermore, such a move would no doubt increase the confrontation of opposed scientific views on national policy.

A third way in which science can be effectively integrated into national policy-making is suggested by a constant theme in contemporary political commentary. This theme is the existence of a great gulf in understanding between the scientist and the non-scientist. C. P. Snow has popularized this notion with his positing of two separate cultures:

that of humanism and that of science.[38] Similarly the present study has been concerned with the fact that although there is an increasingly intimate relationship between the political and scientific cultures, a great gulf continues to exist in the realm of understanding.

In particular this study has shown that although the scientist has made immense contributions to American political life, his background and training as a specialist has often been a source of trained *political* incapacity. At the same time, the failure of political leadership to understand the capabilities and limitations of the scientist has given rise to grave errors in policy-making and policy execution.

Recent writers—aware of the gulf between scientist and public official—have sought to restate the continuing need for the generalist in a period increasingly complex and characterized by the fragmentation of intellectual life into technical specializations. Charles Thayer, writing on diplomacy,[39] Henry Kissinger, on foreign policy formulation,[40] and Don Price, on public administration,[41] have all sought to re-establish the fact that the executors, formulators, and administrators of national policy should be men with "an appropriate mixture of general competence and special knowledge" who will serve "as a useful layer in the pyramid of policy between the peak of political power and the base of science and technology . . . ."[42]

There is a profound need for civil servants and others who can communicate with both the scientist and the political leader. Such men must know the limitations and the competences of the former as well as the problems and responsibilities of the latter. Unfortunately there are few men with these capabilities, and, in the long run, the

[38] C. P. Snow, *The Two Cultures and the Scientific Revolution,* The Cambridge University Press, 1959.

[39] Charles Thayer, *Diplomat,* Harpers, 1959, Chapter 23.

[40] Henry Kissinger, *op.cit.,* Chapter 8.

[41] Don K. Price, *op.cit.,* Chapter 6.

[42] *ibid.,* p. 186.

intellectual brokers between the political and scientific realms will have to be educated to the task.

"What is needed," Dean Price of the Graduate School of Public Administration at Harvard University, writes: "is a corps of men whose liberal education includes an appreciation of the role of science and technology in society and whose scientific education has not been a narrowly technical or vocational one, but has treated science as one of the highest intellectual endeavors of men who also have responsibility as free citizens. The humanities and the social sciences are too often taught in America as narrowly technical subjects. We can hardly found a new generation of administrative generalists on them as they are commonly taught today."[43]

In addition to the administrative-generalist who is familiar with the worlds both of science and of politics, there is need for diplomats and military officers who are knowledgeable in scientific affairs. Men are needed in these professions who know enough science to work effectively with scientists in the formulation and execution of national policy. Similarly there is a need for scientists to understand the nature of politics and to appreciate that there is a vast difference between the scientific and political realms.

Here, then, is another challenge to effective policy-making with respect to nuclear weapons and the other products of the scientific revolution. Society has a responsibility to train, and political leadership has a responsibility to utilize, generalists and specialists who have an intellectual grasp of the new type of technical-political problems which the scientific revolution is bringing forth. Ultimately only in this way can scientific knowledge be integrated into sound national policy-making and execution.

A PROFESSIONAL CODE FOR THE SCIENTIST-ADVISOR

The active participation of scientists in the formulation of national policy toward nuclear weapons has given rise to

[43] *ibid.*, p. 187.

a number of moral and political issues which scientists and society at large must face. These issues pertain to the scientist's rights and responsibilities as an advisor. The failure to recognize and to resolve these issues in large measure underlay the Oppenheimer security trial, the bitterness of intra-scientific conflict over nuclear weapons policy, and the resentment against scientists in various circles. Their resolution is imperative if the mistakes of the past are to be avoided in the future.

In the view of this study these issues of the scientist's rights and responsibilities as an advisor can in part be solved through the evolution of a professional code for the scientist as advisor. Such a code would influence the expectations for the scientist's behavior as advisor just as there is a professional code which prescribes proper behavior for the scientist as researcher. As this study has shown, the fact that the latter code has proved to be inapplicable to the former role has caused many problems within the scientific community since 1945.

The first issue to be met by such a professional code is that of the scientist-advisor's responsibilities and rights with respect to his fellow scientists. This problem has been a source of constant intra-scientific friction since 1945. Throughout this period, scientist-advisors have defined their responsibilities in one of three ways. The first is that the scientist-advisor as representative of the scientific community can follow the advice of Edmund Burke in his letter to the electors of Bristol; that is to say, he speaks only for his own conscience. The second is that the scientist speaks for what he believes to be the majority opinion of scientists. The third is that he is a transmitter to political leadership of all the various views held by scientists.

In 1945 the scientists who advised the Truman Administration on the dropping of the atomic bomb tried to combine an expression of their own opinions on the issue with a survey of all pertinent scientific opinions; nevertheless, those scientists who opposed the dropping of the bomb believed their representatives had failed to present their case adequately. In 1949 the scientists on the General Advisory Committee of the AEC at the time of the hydrogen bomb decision presented to the Administration only their own views on the advisability of a crash program to produce a thermonuclear weapon; rightly or wrongly the scientists in the other camp believed this action of their "representatives" to the AEC to be an act of betrayal. In 1958–60 PSAC has spoken for the majority of the scientific community in its opinion that an inspectable nuclear test ban is technically feasible; the scientific opponents of a test ban (who vigorously deny this) believe that their minority opinion has had inadequate representation at the presidential level.

The failure of each alternative to satisfy important factions within the scientific community has been due to the necessarily political nature of scientific advice. For this reason it is the view of the present study that no scientist-advisor can hope to speak for anything but his own point of view. If another point of view is to be heard it must be presented by an individual who possesses that viewpoint. Thus, it is prudent that the advisory mechanisms contain representatives of many political points of view to be found among scientists on major policy issues. In addition, scientists must appreciate the often parochial nature of their advice. Their conduct as advisors must be based on a realization that they can speak effectively only for their own limited viewpoint.

On the basis of this understanding one can define in part the rights of the scientist-advisor with respect to his fellow scientists. Certainly one right should be the freedom of the scientist to give advice to political leadership without being subjected to accusations of intellectual dishonesty because his interpretation of the facts differs from that of his colleagues. Although objectivity remains a goal for the scientist-advisor—despite the impossibility of its achievement—

the wisdom of his advice ought to be the measure of his usefulness to political leadership.

The second issue with which a professional code must deal is that of the scientist-advisor's responsibilities and rights with respect to his advisee if the advisee is other than the body politic. What, for example, is the responsibility of PSAC to the President if he fails to follow its advice?[44] Does it merely resign itself to the President's decision? Or does it have a higher responsibility to the Constitution and thereby an obligation to take its case to the Congress and the people?

If scientists choose to remain loyal to their advisee and not to appeal their case, one must ask whether or not they can be held responsible for failure to oppose a policy they believe to be wrong. Thus, following President Truman's decision to develop the hydrogen bomb despite its advice, was the GAC negligent in failing to take its case against the bomb to the Congress and, insofar as security regulations permitted, to the public itself? Or ought the responsibility of the advisor to be solely to give his advisee the best advice of which he is capable and then to leave the disposition of that advice solely to the advisee?

There is, as Don Price observes, "no simple answer" to this dilemma for the advisor under the American system. The written Constitution of the United States " . . . puts him [the advisor] in a position of frequently having to decide for himself" whether or not to place loyalty to Constitution above loyalty to an advisee.[45] The realization by sensitive scientist-advisors that such a responsibility has been imposed upon them has become a source of resentment among scientists. They resent the fact that as advisors they must share responsibility with political leadership for decisions made while they believe they lack sufficient authority in the making of the decision itself.

---

[44] This subject has been discussed in detail by Don Price, *ibid.*, pp. 137ff.

[45] *ibid.*, p. 139.

Although it would be wrong for the scientist to aspire to share political authority with elected officials, the fact that he does appear to share certain responsibility with them makes it reasonable to argue that he be granted certain rights as an advisor. At least certain questions concerning his rights ought to be posed: (1) Does the scientist-advisor have the right to initiate advice or must he speak only when spoken to? (2) Ought the scientist-advisor to concern himself with the political, strategic, and moral implications of technical questions or must he refrain from stepping outside his technical competence? (3) Should the scientist-advisor be given broad policy matters on which to give advice or must he be restricted to narrowly prescribed questions?

These issues arise from the various activities of scientists since 1945. For example, in the early 1950's there arose the issue of the breadth of the mandate to be given the scientist as advisor. This occurred with respect to Project Vista which was assigned the responsibility of studying the employment of nuclear weapons in the ground defense of Western Europe. The scientists in the project incurred the wrath of the Air Force by expanding the mandate given them to include an evaluation of the role of the Strategic Air Command in Western strategy. Did the scientists act properly in expanding the breadth of their mandate in the light of their belief that they could not succeed otherwise in their assigned task? The Gray Board majority implied in its interrogation of Oppenheimer that the scientists were wrong.

It is the conviction of the present writer that if the scientist is to share responsibility for decisions, he must be granted certain rights as an advisor. The scientist-advisor must be granted a wide latitude in which to operate. This means an acceptance by the decision-maker of the principle that the advisor has a prerogative to initiate advice, to redefine questions, and to determine his own standards of relevance.[46]

[46] For a brief discussion of this matter see Henry Kissinger, *op.cit.*, pp. 348–54.

But, in turn, the advisor must realize that his utility rests ultimately upon his ability to help the decision-maker solve the problems which confront him.

It is certainly valid to question whether or not the scientist has the right consciously to advise the public or its representatives on non-technical matters as the GAC scientists did in the 1949 hydrogen bomb controversy. The view of this study is that scientists ought to have a right to advise political leaders on non-technical matters; it does not necessarily follow that because scientists are primarily technical experts they are incompetent to advise on political or strategic matters.

If, however, scientists choose to exercise the right—or perhaps, more appropriately, the privilege—of advising political leaders on non-technical matters, they must then appreciate the fact that they thus assume a number of responsibilities. In the first place, when giving political advice, scientists have an obligation to prepare themselves as thoroughly with respect to the political and social aspects of the problem under consideration as on the scientific and technical aspects of a problem within their own scientific discipline. Only to the degree that natural scientists do educate themselves in social and political matters ought they to expect others to hearken to their political advice.

A second responsibility of scientists giving political advice, especially to the public, is to attempt to avoid or, at least, to moderate appeals to passion. Certainly scientists must themselves set a standard of rationality in this troubled age and discontinue any attempts to "save mankind by frightening men to rationality"; it is doubtful that emotional appeals can do anything other than inhibit intelligent thought on how to survive and remain free in the atomic age. Furthermore—and fortunately this criticism applies only to a minority of scientists—scientists must overcome an anti-intellectualism which argues that any attempt to study the problem of thermonuclear war scientifically is immoral because it contributes to an acceptance of that type of war.

Thirdly and most importantly, if scientists are to participate in government and to advise on policy matters, they must appreciate the possibly unpleasant aspects of their new role. The prestige of governmental service is accompanied by the risk of criticism and downfall if one's advice proves unpopular or results in policy errors. The scientist is no more immune from the ravages of the political system than are other public servants such as Harold Stassen and Lewis Strauss. The closer he gets to having a determining influence on policy, the more he will be held responsible for his acts and subject to political dismissal. The scientific community therefore cannot expect its members to share political responsibility and prestige without on occasion having to suffer defeat. Scientists must recognize the right of the political executive to choose his scientific advisors for political sympathy as well as for their scientific competence. On the other hand, scientist-advisors have a right to expect that they will not be punished for their policy views by charges that reflect on their character.

In summary, the object of a professional code for scientific advisors would be the creation of an atmosphere where wisdom could flourish. The United States would certainly be ill-advised to circumscribe severely the participation of American scientists in national decision-making. Yet at the same time the scientists who become advisors assume certain responsibilities with respect to fellow scientists, their advisees, and above all, to the body politic. Their value as advisors resides in their capacity and in that of political leadership to attain the proper balance among their own rights and these responsibilities.

Conclusion

Since World War II, American scientists have contributed to the search for the solution to the nuclear dilemma that has faced the United States. The major challenge to American security policy has been that of choosing between disarmament and deterrence as the means of achieving strategic stability. With the hydrogen bomb decision in 1949, the United States began to foreclose the choice of disarmament, but without actually choosing strategic deterrence as its goal. The reorientation of American policy in 1955 toward arms control rather than disarmament and the acceleration of the missile program were significant moves toward strategic deterrence.[47] Yet on September 25, 1961, the United States announced its acceptance of the goal of complete and general disarmament; this announcement indicates that at least on the verbal level the nuclear dilemma still exists.

In all of these decisions centering around the problem of America's nuclear dilemma scientists have had an important part to play. While this study has frequently been critical of these scientists, it has been so because the emergence of scientists into the mainstream of American political life is one of the great events of American history. Not because the political conduct of scientists has been especially censurable, although it has been that at times, but because scientists have become men of power and because in a democracy power must be constantly and responsibly evaluated, there is a need for both scientists and non-scientists to understand the scientist as a political animal.

Since 1945 American political leadership has in general depended upon those scientists who through a sense of social responsibility have been impelled to participate in national policy making. The need for increased numbers of advisors, the changing nature of competences required, and the gradual retirement of the present generation of advisors will necessitate a more conscious effort to recruit scientists into government. Political leadership will therefore be required to recruit, train, and nourish those scientists with broadened interests and tempered judgments of whom Paul Nitze has written.[48]

---

[47] An excellent history of the political struggle over defense policy is Samuel Huntington, *The Common Defense*, Columbia University Press, 1961.

[48] Paul Nitze, *op.cit.*

Furthermore, the brief history of American nuclear policy does not encourage one to believe that dependence upon a restricted intellectual elite is a sound practice in a democracy. For this reason, the admonition to the American people given by Robert Oppenheimer shortly after President Truman's decision to attempt the development of the hydrogen bomb is still sage advice: "The decision to seek or not to seek international control of the A-bomb, the decision to try to make or not to make the H-bomb, are issues, rooted in complex technical matters, that nevertheless touch the very basis of our morality. There is grave danger for us in that these decisions have been taken on the basis of facts held secret. This is not because the men who must contribute to the decisions, or must make them, are lacking in wisdom; it is because wisdom itself cannot flourish, nor even truth be determined, without the give and take of debate or criticism. The relevant facts could be of little help to an enemy; yet they are indispensable for an understanding of questions of policy. If we are wholly guided by fear, we shall fail in this time of crisis. The answer to fear cannot always lie in the dissipation of the causes of fear; sometimes it lies in courage."[49]

[49] This statement was made February 12, 1950 on Mrs. Franklin Roosevelt's program, *Round Table*. See "Fateful Decision," *Bull. Atom. Sci.*, Vol. 6, No. 3, March 1950, p. 75.

INDEX

Acheson, Dean, 53, 95, 112
Acheson-Lilienthal Proposals, 53ff.
  See also Baruch Plan
AEC, *see* Atomic Energy Commission
Alvarez, Luis, 83
American Association for the Advancement of Science, 216
Anderson, Clinton, 249
anti-missile missile, 254, 283
arms control, 37, 164, 233, 259, 310, 312, 319, 321–23
Arms Control and Disarmament Agency, 332
atmospheric nuclear tests, 186, 191. *See also* nuclear test ban, Geneva System
atomic bomb; effect on scientists' attitudes, 25, 28, 39, 49–50; decision to drop, 27, 45–46, 47; effect on American security, 40–41, 70, 77, 118
"atomic club," 11, 149
Atomic Development Authority, 53ff, 103
Atomic Energy Commission, 3, 12, 50–51, 139. *See also* General Advisory Committee
atomic energy, international control of, 39ff, 65, 131, 164. *See also* Baruch Plan; open world, need for
atomic energy legislation, 12, 129, 173
Atomic Energy Panel of the Research and Development Board of the Department of Defense, 114
atomic energy, peaceful uses of, *see* Atoms for Peace
atomic espionage, 80
atomic tests, 124–25, 174. *See also* Bikini, Bravo, Hardtack, Rainier, and Operations Greenhouse, Ivy, and Sandstone

"Atoms for Peace" proposal, 128, 152, 296

B-36 controversy, 89
Bacher, Robert, 115, 185
"balance of terror," 145
Baldwin, Hanson, 81
Baruch, Bernard, 53
Baruch Plan, 31, 35, 37, 52ff, 60–62, 65
Berkner, Lloyd, 230
Berkner Committee, *see* Panel on Seismic Improvement
Bethe, Hans, 18, 35, 69, 290–91; on social responsibility of scientists 24; and hydrogen bomb, 82, 86, 100–101, 121; on feasibility of a nuclear test ban, 159, 178, 262, 268ff, 324; political attack on, 196; on Hardtack data, 228; on decoupling theory, 235–36, 301; at Technical Working Group II, 240; reaction to Soviet proposal of March 19, 1960, 249–51; evaluation of Soviet nuclear progress (1961), 254; on resumption of atmospheric tests, 255; on lesson of test ban negotiations, 257
Bethe Panel (1958), 179–80, 207; membership of, 179
Bethe Panel (1961), 254
"big-hole" theory, *see* decoupling
Bikini tests, 124
Bohr Memorandum, 42–44, 102, 330
Bohr, Niels, 42, 52
Borden, William L., 132
Bradbury, Norris, 5–6
Bradley, Omar, 88
Bravo test, 122, 138
Brennan, Donald, 293
Bridgman, Percy, 26–27
Brown, Harold, 204ff
Brown, Harrison, 273, 301